T0212788

Lecture Notes in Computer Science 9025

Commenced Publication in 1973
Founding and Former Series Editors:
Gerhard Goos, Juris Hartmanis, and Jan van Leeuwen

Editorial Board

David Hutchison
Lancaster University, Lancaster, UK
Takeo Kanade
Carnegie Mellon University, Pittsburgh, PA, USA
Josef Kittler
University of Surrey, Guildford, UK
Jon M. Kleinberg
Cornell University, Ithaca, NY, USA
Friedemann Mattern
ETH Zurich, Zürich, Switzerland
John C. Mitchell
Stanford University, Stanford, CA, USA
Moni Naor
Weizmann Institute of Science, Rehovot, Israel
C. Pandu Rangan
Indian Institute of Technology, Madras, India
Bernhard Steffen
TU Dortmund University, Dortmund, Germany
Demetri Terzopoulos
University of California, Los Angeles, CA, USA
Doug Tygar
University of California, Berkeley, CA, USA
Gerhard Weikum
Max Planck Institute for Informatics, Saarbrücken, Germany

More information about this series at http://www.springer.com/series/7407

Penousal Machado · Malcolm I. Heywood
James McDermott · Mauro Castelli
Pablo García-Sánchez · Paolo Burelli
Sebastian Risi · Kevin Sim (Eds.)

Genetic Programming

18th European Conference, EuroGP 2015
Copenhagen, Denmark, April 8–10, 2015
Proceedings

 Springer

Editors

Penousal Machado
Universidade de Coimbra
Coimbra
Portugal

Malcolm I. Heywood
Dalhousie University
Halifax, NS
Canada

James McDermott
University College Dublin
Dublin
Ireland

Mauro Castelli
Universidade Nova de Lisboa
Lisboa
Portugal

Pablo García-Sánchez
Universidad de Granada
Granada
Spain

Paolo Burelli
Aalborg University
Copenhagen
Denmark

Sebastian Risi
IT University of Copenhagen
Copenhagen
Denmark

Kevin Sim
Edinburgh Napier University
Edinburgh
UK

ISSN 0302-9743   ISSN 1611-3349   (electronic)
Lecture Notes in Computer Science
ISBN 978-3-319-16500-4   ISBN 978-3-319-16501-1   (eBook)
DOI 10.1007/978-3-319-16501-1

Library of Congress Control Number: 2015933505

LNCS Sublibrary: SL1 – Theoretical Computer Science and General Issues

Springer Cham Heidelberg New York Dordrecht London
© Springer International Publishing Switzerland 2015
This work is subject to copyright. All rights are reserved by the Publisher, whether the whole or part of the material is concerned, specifically the rights of translation, reprinting, reuse of illustrations, recitation, broadcasting, reproduction on microfilms or in any other physical way, and transmission or information storage and retrieval, electronic adaptation, computer software, or by similar or dissimilar methodology now known or hereafter developed.
The use of general descriptive names, registered names, trademarks, service marks, etc. in this publication does not imply, even in the absence of a specific statement, that such names are exempt from the relevant protective laws and regulations and therefore free for general use.
The publisher, the authors and the editors are safe to assume that the advice and information in this book are believed to be true and accurate at the date of publication. Neither the publisher nor the authors or the editors give a warranty, express or implied, with respect to the material contained herein or for any errors or omissions that may have been made.

Cover illustration: Designed by Mauro Castelli, ISEGI, Universidade Nova de Lisboa, Portugal

Printed on acid-free paper

Springer International Publishing AG Switzerland is part of Springer Science+Business Media
(www.springer.com)

Preface

The 18th European Conference on Genetic Programming (EuroGP) took place during April 8–10, 2015. Copenhagen, Denmark was the setting, and the *Nationalmuseet* was the venue. EuroGP is the only conference exclusively devoted to the evolutionary generation of computer programs and attracts scholars from all over the world. The maturity of the event is in part reflected by the fact that 'Google Scholar' now lists EuroGP as one of the top 20 venues in Evolutionary Computation with an h5-index and h5-median of 15 and 18 respectively.[1] Collectively, over 10,000 articles now appear in the online GP bibliography maintained by William B. Langdon.[2]

The unique character of genetic programming has been recognized from its very beginning. EuroGP has had an essential impact on the success of the field, by serving as an important forum for expressing new ideas, meeting fellow researchers, and starting collaborations. Indeed, EuroGP represents the single largest venue at which genetic programming results are published. Many success stories have been witnessed by the 18 editions of EuroGP. To date, genetic programming is essentially the only approach that has demonstrated the ability to automatically generate, repair, and improve computer code in a wide variety of problem areas. It is also one of the leading methodologies that can be used to 'automate' science, helping researchers to induce hidden complex models from observed phenomena. Furthermore, genetic programming has been applied to many problems of practical significance, and has produced human-competitive solutions.

EuroGP 2015 received 36 submissions from 21 different countries across five continents. The papers underwent a rigorous double-blind peer-review process, each being reviewed by at least three members of the international Program Committee from 23 countries. The selection process resulted in this volume, with 12 papers accepted for oral presentation (33.3 % acceptance rate) and 6 for poster presentation (50 % global acceptance rate for talks and posters combined). The wide range of topics in this volume reflects the current state of research in the field. Thus, we see topics as diverse as semantic methods, recursive programs, grammatical methods, coevolution, Cartesian GP, feature selection, initialization procedures, ensemble methods, and search objectives; and applications including text processing, cryptography, numerical modeling, software parallelization, creation and optimization of circuits, multi-class classification, scheduling, and artificial intelligence.

Together with three other colocated evolutionary computation conferences (Evo-COP 2015, EvoMUSART 2015, and EvoApplications 2015), EuroGP 2015 was part of the Evo* 2015 event. This meeting could not have taken place without the help of many people.

---

[1] http://scholar.google.com/citations?view_op=top_venues&vq=eng_evolutionarycomputation.

[2] http://liinwww.ira.uka.de/bibliography/Ai/genetic.programming.html.

First to be thanked is the great community of researchers and practitioners who contributed to the conference by both submitting their work and reviewing others' as part of the Program Committee. Their hard work, in evolutionary terms, provided both variation and selection, without which progress in the field would not be possible!

The papers were submitted, reviewed, and selected using the MyReview conference management software. We are sincerely grateful to Marc Schoenauer of Inria, France, for his great assistance in providing, hosting, and managing the software.

We would like to thank the local organising team: Paolo Burelli from the Faculty of Engineering and Science, Aalborg University and Sebastian Risi, codirector of the Robotics, Evolution and Art Laboratory at the IT University of Copenhagen.

We thank Kevin Sim from the Institute for Informatics & Digital Information, Edinburgh Napier University for creating and maintaining the official Evo* 2014 website, and Pablo Garía-Sánchez (Universidad de Granada, Spain) and Mauro Castelli (Universidade Nova de Lisboa, Portugal) for being responsible for Evo* 2014 publicity.

We would also like to express our sincerest gratitude to our invited speakers, who gave inspiring keynote talks: Professor Paulien Hogeweg of the Bioinformatics group, Utrecht University, The Netherlands, and Dr. Pierre-Yves Oudeyer, Research Director at Inria, Paris, France.

We especially want to express our genuine gratitude to Jennifer Willies of the Institute for Informatics and Digital Innovation at Edinburgh Napier University, UK. Her dedicated and continued involvement in Evo* since 1998 has been and remains essential for building the image, status, and unique atmosphere of this series of events.

April 2015

Penousal Machado
Malcolm I. Heywood
James McDermott
Mauro Castelli
Pablo García-Sánchez
Paolo Burelli
Kevin Sim

Organization

Administrative details were handled by Jennifer Willies, Institute for Informatics and Digital Innovation, Edinburgh Napier University, Scotland, UK.

Organizing Committee

Program Co-chairs

Penousal Machado   University of Coimbra, Portugal
Malcolm I. Heywood   Dalhousie University, Canada

Publication Chair

James McDermott   University College Dublin, Ireland

Publicity Chairs

Mauro Castelli   Universidade Nova de Lisboa, Portugal
Pablo García-Sánchez   Universidad de Granada, Spain

Local Chairs

Paolo Burelli   Aalborg University, Denmark
Sebastian Risi   IT University of Copenhagen, Denmark

Webmaster

Kevin Sim   Edinburgh Napier University, UK

Program Committee

Alexandros Agapitos	University College Dublin, Ireland
Lee Altenberg	University of Hawaii at Manoa, USA
R. Muhammad Atif Azad	University of Limerick, Ireland
Ignacio Arnaldo	Massachusetts Institute of Technology, USA
Douglas Augusto	LNCC/UFJF, Brazil
Wolfgang Banzhaf	Memorial University of Newfoundland, Canada
Mohamed Bahy Bader	University of Portsmouth, UK
Helio Barbosa	LNCC/UFJF, Brazil
Heder Bernardino	LNCC/UFJF, Brazil
Anthony Brabazon	University College Dublin, Ireland
Nicolas Bredeche	Université Pierre et Marie Curie, France
Stefano Cagnoni	University of Parma, Italy
Ernesto Costa	University of Coimbra, Portugal
Luis Da Costa	Université Paris-Sud XI, France
Antonio Della Cioppa	University of Salerno, Italy

Federico Divina	Pablo de Olavide University, Spain
Marc Ebner	Ernst-Moritz-Arndt Universität Greifswald, Germany
Aniko Ekart	Aston University, UK
Francisco Fernández de Vega	Universidad de Extremadura, Spain
Gianluigi Folino	ICAR-CNR, Italy
James A. Foster	University of Idaho, USA
Christian Gagné	Université Laval, Québec, Canada
Steven Gustafson	GE Global Research, USA
Jin-Kao Hao	LERIA, University of Angers, France
Inman Harvey	University of Sussex, UK
Erik Hemberg	Massachusetts Institute of Technology, USA
Malcolm I. Heywood	Dalhousie University, Canada
Ting Hu	Dartmouth College, USA
David Jackson	University of Liverpool, UK
Colin Johnson	University of Kent, UK
Tatiana Kalganova	Brunel University, UK
Ahmed Kattan	Umm Al-Qura University, Saudi Arabia
Graham Kendall	University of Nottingham, UK
Michael Korns	Korns Associates, USA
Jan Koutnik	IDSIA Dalle Molle Institute for Artificial Intelligence, Switzerland
Krzysztof Krawiec	Poznan University of Technology, Poland
Jiri Kubalik	Czech Technical University in Prague, Czech Republic
William B. Langdon	University College London, UK
Kwong Sak Leung	The Chinese University of Hong Kong, China
John Levine	University of Strathclyde, UK
Evelyne Lutton	Inria, France
Penousal Machado	University of Coimbra, Portugal
Radek Matousek	Brno University of Technology, Czech Republic
James McDermott	University College Dublin, Ireland
Andrew McIntyre	Dalhousie University, Canada
Bob McKay	Seoul National University, Korea
Jorn Mehnen	Cranfield University, UK
Julian Miller	University of York, UK
Alberto Moraglio	University of Exeter, UK
Xuan Hoai Nguyen	Hanoi University, Vietnam
Miguel Nicolau	University College Dublin, Ireland
Julio Cesar Nievola	Pontificia Universidade Católica do Paraná, Brazil
Michael O'Neill	University College Dublin, Ireland
Una-May O'Reilly	Massachusetts Institute of Technology, USA
Fernando Otero	University of Kent, UK
Ender Ozcan	University of Nottingham, UK
Andrew J. Parkes	University of Nottingham, UK
Gisele Pappa	Federal University of Minas Gerais, Brazil

Tomasz Pawlak	Poznan University of Technology, Poland
Clara Pizzuti	Institute for High Performance Computing and Networking, Italy
Thomas Ray	University of Oklahoma, USA
Peter Rockett	University of Sheffield, UK
Denis Robilliard	Université Lille Nord de France, France
Conor Ryan	University of Limerick, Ireland
Marc Schoenauer	Inria, France
Lukas Sekanina	Brno University of Technology, Czech Republic
Yin Shan	Medicare, Australia
Sara Silva	INESC-ID Lisboa, Portugal
Moshe Sipper	Ben-Gurion University, Israel
Alexei N. Skurikhin	Los Alamos National Laboratory, USA
Terence Soule	University of Idaho, USA
Lee Spector	Hampshire College, USA
Ivan Tanev	Doshisha University, Japan
Ernesto Tarantino	ICAR-CNR, Italy
Jorge Tavares	Microsoft, Germany
Leonardo Trujillo	Instituto Tecnológico de Tijuana, Mexico
Leonardo Vanneschi	Universidade Nova de Lisboa, Portugal, and University of Milano-Bicocca, Italy
Man Leung Wong	Lingnan University, Hong Kong
Lidia Yamamoto	University of Strasbourg, France
Mengjie Zhang	Victoria University of Wellington, New Zealand

Contents

Posters

Regular Papers

The Effect of Distinct Geometric Semantic Crossover Operators in Regression Problems

Julio Albinati[1]([✉]), Gisele L. Pappa[1],
Fernando E.B. Otero[2], and Luiz Otávio V.B. Oliveira[1,3]

[1] Universidade Federal de Minas Gerais, Belo Horizonte, Brazil
{jalbinati,glpappa,luizvbo}@dcc.ufmg.br
[2] Chatham Maritime, University of Kent, Kent, UK
F.E.B.Otero@kent.ac.uk
[3] Instituto Federal Do Sul de Minas Gerais, Poços de Caldas, Brazil

Abstract. This paper investigates the impact of geometric semantic crossover operators in a wide range of symbolic regression problems. First, it analyses the impact of using Manhattan and Euclidean distance geometric semantic crossovers in the learning process. Then, it proposes two strategies to numerically optimize the crossover mask based on mathematical properties of these operators, instead of simply generating them randomly. An experimental analysis comparing geometric semantic crossovers using Euclidean and Manhattan distances and the proposed strategies is performed in a test bed of twenty datasets. The results show that the use of different distance functions in the semantic geometric crossover has little impact on the test error, and that our optimized crossover masks yield slightly better results. For SGP practitioners, we suggest the use of the semantic crossover based on the Euclidean distance, as it achieved similar results to those obtained by more complex operators.

Keywords: Semantic genetic programming · Crossover · Crossover mask optimization

1 Introduction

The development of methods that take the semantics of the solutions being evolved into account is a trend in the genetic programming community, with special attention given to methods based on geometric semantic crossover operators [8,12]. The main reason for researchers interest in geometric semantic crossover is that, by manipulating directly the semantics of solutions, it behaves in a much more controlled way since the semantic impact of operators can be easily bounded. It also has interesting properties regarding control of overfitting, establishing upper bounds in test error. Furthermore, the fitness landscapes induced by semantic operators are usually much simpler than regular landscapes, making optimization easier.

© Springer International Publishing Switzerland 2015
P. Machado et al. (Eds.): EuroGP 2015, LNCS 9025, pp. 3–15, 2015.
DOI: 10.1007/978-3-319-16501-1_1

This paper is particularly interested in the impact of geometric semantic crossover operators into symbolic regression problems. Given a set of inputs I and their respectively expected outputs O, the semantics of a function f being evolved can be indirectly assessed by a quality measure, such as the error rate, which calculates the differences between the expected outputs O and the obtained outputs $O' = f(I)$. As different functions can map to the same value of error, we can say this error measure performs a kind of syntactic–semantic (or genotype-phenotype) mapping.

Considering that the semantics of a solution can be represented by its output vector O', different ways of measuring the semantic distance between two functions have been proposed. In the case of symbolic regression, as the outputs generated return real values, the Manhattan and Euclidean distances are appropriate functions for measuring error. These distance metrics can then be used to measure the semantic distance between pairs of individuals.

Geometric semantic operators were defined to work into different spaces of functions defined by the distance metrics previously described. In this way, the geometric crossover is a function of the semantics of their parents. Given two real-valued functions f_1 and f_2, a geometric semantic crossover returns a third real-valued function f_3 representing the convex combination of the parents. The offspring is obtained by multiplying f_1 and f_2 by a *crossover mask*, which can be represented by a constant, in the case of the Euclidean distance or a function (e.g. logistic [13]), in the case of the Manhattan distance.

This paper investigates the impact of using Manhattan and Euclidean distance on geometric semantic crossovers in the learning process and proposes two strategies to numerically optimize the crossover mask based on mathematical properties of these operators, instead of simply generating them randomly. We present an experimental analysis in which the different distance metric crossovers and the proposed strategies are compared on a test bed composed of twenty datasets with distinct properties from both real and synthetic domains.

The remainder of this paper is organised as follows. Section 2 presents an overview of the methods that incorporate semantic awareness into GP. Section 3 introduces the two new optimization strategies applied to the crossover masks, followed by the experimental analysis in the test bed in Sect. 4. Finally, conclusions and perspectives of future work are presented in Sect. 5.

2   Related Work

The study of programs or individuals semantics in GP has been developed mostly in the last five years [12]. In [12], the authors divide semantic-aware techniques into three groups: *diversity methods*, *indirect semantic methods* and *direct semantic methods*.

Diversity methods were the first proposed, aiming to preserve or reinject diversity throughout evolution. Although methods aiming at GP diversity are not new, they usually considered only syntactic diversity. In [4], in turn, the authors studied the impact of semantic diversity during population initialization, showing that greater diversity leads to improved results. Indirect semantic methods, on the other hand, use regular GP operator, but only accept individuals if they respect some semantic-related criteria, such as their semantic difference to their parents [3] or to a geometric (semantically intermediate) individual of their parents [6]. Different versions of methods based on this approach were subsequently proposed [9,10], and although they led to improved results over traditional crossover operators, they are *trial-and-error* techniques, without any guarantees that solutions respecting the criteria established will be actually generated.

In contrast to indirect methods, direct semantic methods use operators specifically designed to operate in a semantic level, and are our subject of interest. In [8], the authors proposed geometric semantic crossover and mutation operators for three domains: boolean, categorical and real-valued. The geometric semantic crossover operator follows Definition 1, which is essentially a convex combination of two previously generated solutions. The function that combines the solutions, $c(\mathbf{x})$, is the *crossover mask*.

Definition 1. *Let $\mathcal{F}$ be the set of functions mapping instances to real numbers and $f_1(\boldsymbol{x}), f_2(\boldsymbol{x}) \in \mathcal{F}$ be two previously generated solutions. Then $XO : \mathcal{F} \times \mathcal{F} \to \mathcal{F}$ is called a semantic geometric crossover (for real-valued functions) if $XO(f_1(\boldsymbol{x}), f_2(\boldsymbol{x})) = c(\boldsymbol{x}) \cdot f_1(\boldsymbol{x}) + (1 - c(\boldsymbol{x})) \cdot f_2(\boldsymbol{x})$, where $c(\boldsymbol{x})$ outputs values in the interval $[0, 1]$.*

Note that for the crossover operator, if $c(\mathbf{x}) = \beta$ for all $\mathbf{x}$, then its geometric properties will be related to the Euclidean distance in the semantic search space. However, if $c(\mathbf{x})$ is allowed to output distinct values for distinct instances, then its geometric properties will be related to the Manhattan distance.

Since the crossover operator performs a convex combination, a solution $f_3(\mathbf{x})$ generated by this operator will be semantically intermediate of its parents $f_1(\mathbf{x})$ and $f_2(\mathbf{x})$: $dist(f_1, f_2) = dist(f_1, f_3) + dist(f_3, f_2)$, where $dist$ may be the Euclidean or Manhattan distance in the semantic search space depending on the choice of crossover mask. For the Euclidean distance, this fact lead to the property that the error committed by $f_3(\mathbf{x})$ is upper bounded by the error of the worst of its parents. More interestingly, this property holds for both training and test sets, thus being useful for controlling overfitting [8,11].

Successive applications of the geometric semantic crossover, however, may lead to an exponential growth in the size of solutions, as pointed out in [8]. In fact, the number of nodes of a tree T_3 obtained as the crossover of two other trees, T_1 and T_2, is greater than the number of nodes of T_1 and T_2 altogether. Although it is possible to simplify the functions represented by such trees, this would lead to a large computational effort. In [5], the authors proposed an efficient way of dealing with this problem by avoiding replicating subtrees that are part of more than a solution. They also suggested the usage of a sigmoid (logistic) function $c(\mathbf{x}) = (1 + e^{-r(\mathbf{x})})^{-1}$ ($r(\mathbf{x})$ being a randomly generated function), which correctly outputs values in the required interval.

3 Optimized Semantic Crossover Operators

This section proposes two new versions of the semantic crossover operator showed in Definition 1. These new versions were generated by optimizing the crossover masks instead of randomly generating them.

3.1 Optimized Convex Geometric Semantic Crossover Operator

The first proposed operator was generated by finding the value of the crossover mask that leads to the minimum training error. As we show in this section, this new operator has an interesting property: it is *non-degenerative*, strengthening the convex property regarding training error. While the convex property states that the error of the function being generated will never be larger than the worst of its parents, we can now state that the error of the function being generated will never be larger than the best of its parents when considering training error, as showed below.

Suppose that the crossover mask $c(\mathbf{x})$ is a single constant, i.e., $c(\mathbf{x}) = \beta$ for all $\mathbf{x}$ (and the crossover is based on the Euclidean distance in the semantic space). Let $f_1(\mathbf{x})$, $f_2(\mathbf{x})$ be two previously generated solutions, and $f_3(\mathbf{x}) = XO(f_1(\mathbf{x}), f_2(\mathbf{x}))$. Then, the sum of squared errors (SSE) of f_3 can be expressed in terms of β:

$$SSE(\beta) = \sum_{i=1}^{n} [y_i - \beta \cdot f_1(\mathbf{x}_i) - (1 - \beta) \cdot f_2(\mathbf{x}_i)]^2 \qquad (1)$$

where training data is represented as sequence of pairs $\{(\mathbf{x}_i, y_i)\}_{i=1}^{n}$.

Since $SSE(\beta)$ is continuous, we can calculate the derivative of Eq. 1 and equals it to zero, finding

$$\beta^* = \frac{\sum_{i=1}^{n} [y_i - f_2(\mathbf{x}_i)][f_2(\mathbf{x}_i) - f_1(\mathbf{x}_i)]}{\sum_{i=1}^{n} [f_1(\mathbf{x}_i) - f_2(\mathbf{x}_2)]^2} \qquad (2)$$

such that it minimizes the error of $f_3(\mathbf{x})$, as shown in Proposition 1. Note that calculating the optimized coefficient can be done in $O(n)$, the same time required for computing the semantic of the offspring. Therefore, the optimization of coefficients does not change the asymptotic complexity of SGP.

However, β^* as computed in Eq. 2 may not fall in the $[0, 1]$ interval. This will happen whenever any convex combination of f_1 and f_2 is worse than f_1 and f_2. To enforce the interval constraint, we use a distinct value β^{**} such that

$$\beta^{**} = max(min(1, \beta^*), 0) \qquad (3)$$

Proposition 1. *The argument β^{**} as expressed in Eq. 3 minimizes the error function (Eq. 1) while respecting the interval constraint.*

Proof. Since $\lim_{\beta \to c} SSE(\beta) = SSE(c), \forall c \in \mathbb{R}$, $SSE(\beta)$ is a continuous function in $\mathbb{R}$ and we can compute its derivate with relation to β.

$$\frac{\delta SSE}{\delta \beta} = 2 \cdot \sum_{i=1}^{n} [y_i - f_2(\mathbf{x}_i)][f_2(\mathbf{x}_i) - f_1(\mathbf{x}_i)] + \beta \cdot [f_2(\mathbf{x}_i) - f_1(\mathbf{x}_i)]^2$$

By making the derivative equals to zero, we find a (local) minimum or maximum point.

$$\beta^* = \frac{\sum_{i=1}^{n} [y_i - f_2(\mathbf{x}_i)][f_1(\mathbf{x}_i) - f_2(\mathbf{x}_i)]}{\sum_{i=1}^{n} [f_2(\mathbf{x}_i) - f_1(\mathbf{x}_i)]^2}$$

We now need to show that β^* is a minimization point. We compute the second derivative of $SSE(\beta)$ with relation to β.

$$\frac{\delta^2 SSE(\beta)}{\delta \beta^2} = \sum_{i=1}^{n} [f_2(\mathbf{x}_i) - f_1(\mathbf{x}_i)]^2$$
$$\geq 0$$

Since the second derivative obtained is always non-negative, we prove that β^* is a minimization point and $SSE(\beta)$ is convex. Suppose now that $\beta^* > 1$. Then, $\beta^{**} = 1$ and it minimizes $SSE(\beta)$ while being in the interval $[0, 1]$, since $SSE(\beta)$ is convex and β^{**} is the closest point to β^* in the interval. An analogue reasoning implies that if $\beta^* < 0$, $\beta^{**} = 0$ minimizes the error function while being in the required interval. Finally, if $0 \leq \beta^* \leq 1$, then $\beta^{**} = \beta^*$ and is also a minimization point in the required interval. Thus, we prove that β^{**} minimizes $SSE(\beta)$ while respecting the interval constraint.

As β is optimized in the closed interval $[0, 1]$, if $0 < \beta^{**} < 1$, then $SSE(\beta^{**}) \leq SSE(1)$ and $SSE(\beta^{**}) \leq SSE(0)$. Otherwise, β^{**} would be 1 or 0 and the best of the two functions used in crossover would be simply replicated (see Fig. 1). This shows that the error of the function being generated will never be larger than the best of its parents. Regarding test error, the geometric property is still valid, since this modified version is still a convex combinations of functions.

3.2 Optimized Non-Convex Geometric Semantic Crossover Operator

The convex geometric semantic crossover operator can be very constrained: the fact that it can only generate solutions semantically intermediate of the functions being used for crossover implies that the performance can be strongly determined by the initial population. In order to build a more flexible operator, we propose the following non-convex crossover operator, based on linear combinations (not necessarily convex). However, this means increasing the risk of overfitting, as we now do not have any guarantees regarding test error.

$$XO(f_1(\mathbf{x}_i), f_2(\mathbf{x}_i)) = \beta_1 \cdot f_1(\mathbf{x}_i) + \beta_2 \cdot f_2(\mathbf{x}_i) \tag{4}$$

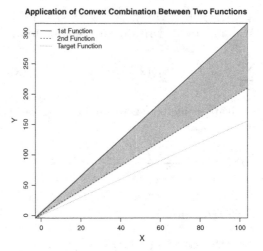

Fig. 1. Application of the convex semantic geometric crossover over functions $Y = 3X + 5$ and $Y = 2X + 2$, and the target function $Y = 1.5X$. The gray area represents possible convex combination of the two functions. Note, however, that the target function is outside of the gray area, meaning that any convex combination is worse (or equal) than the second function.

Again, we can express the error of a function generated through Eq. 4 in terms of β_1 and β_2.

$$SSE(\beta_1, \beta_2) = \sum_{i=1}^{n} [y_i - \beta_1 \cdot f_1(\mathbf{x}_i) - \beta_2 \cdot f_2(\mathbf{x}_i)]^2 \tag{5}$$

Since $SSE(\beta_1, \beta_2)$ is continuous in $\mathbb{R}^2$, we can use the same strategy presented in the previous subsection to find β_1^* and β_2^* that minimizes Eq. 5. Let F be a n-by-2 matrix where $F_{ij} = f_j(\mathbf{x}_i)$ and Y be a column vector of length n containing the target values of each training instance. Then

$$\begin{pmatrix} \beta_1^* \\ \beta_2^* \end{pmatrix} = (F^t F)^{-1} F^t Y \tag{6}$$

Note that similar approaches have been already proposed. In [2], the authors propose to linearly combine *subexpressions* of programs to re-interpret their semantics. In this work, however, we propose to apply a linear combination of two *distinct* programs.

Proposition 2. *The arguments β_1^* and β_2^* as expressed in Eq. 6 minimize the error function (Eq. 5).*

Proof. Since $\lim_{(\beta_1, \beta_2) \to (c_1, c_2)} SSE(\beta_1, \beta_2) = SSE(c_1, c_2)$ for an arbitrary pair $(c_1, c_2) \in \mathbb{R}^2$, $SSE(\beta_1, \beta_2)$ is continuous in $\mathbb{R}^2$ and we can compute its derivative with relation to β_1 and β_2.

$$\frac{\delta SSE}{\delta \beta_1} = \sum_{i=1}^{n} -2[y_i \cdot f_1(\mathbf{x}_i) - \beta_1 \cdot f_1(\mathbf{x}_i)^2 - \beta_2 \cdot f_1(\mathbf{x}_i) \cdot f_2(\mathbf{x}_i)]$$

$$\frac{\delta SSE}{\delta \beta_2} = \sum_{i=1}^{n} -2[y_i \cdot f_2(\mathbf{x}_i) - \beta_1 \cdot f_1(\mathbf{x}_i) \cdot f_2(\mathbf{x}_i) - \beta_2 \cdot f_2(\mathbf{x}_i)^2]$$

Letting F be a n-by-2 matrix where $F_{ij} = f_j(\mathbf{x}_i)$ and Y be a column vector of length n containing the target values for each training instance. By making the derivatives above equal to zero, we arrive in the following matrix formulation:

$$(F^t F) \begin{pmatrix} \beta_1^* \\ \beta_2^* \end{pmatrix} = F^t Y \Rightarrow \begin{pmatrix} \beta_1^* \\ \beta_2^* \end{pmatrix} = (F^t F)^{-1} F^t Y$$

Therefore, we only need to show that (β_1^*, β_2^*) consists of a minimization (and not maximization) point. For that, we will need second order derivatives of SSE.

$$\frac{\delta^2 SSE}{\delta \beta_1^2} = 2 \sum_{i=1}^{n} f_1(\mathbf{x}_i)^2$$

$$\frac{\delta^2 SSE}{\delta \beta_2^2} = 2 \sum_{i=1}^{n} f_2(\mathbf{x}_i)^2$$

$$\frac{\delta^2 SSE}{\delta \beta_1 \delta \beta_2} = 2 \sum_{i=1}^{n} f_1(\mathbf{x}_i) \cdot f_2(\mathbf{x}_i)$$

Since

$$D = \frac{\delta^2 SSE}{\delta \beta_1^2}(\beta_1^*, \beta_2^*) \cdot \frac{\delta^2 SSE}{\delta \beta_2^2}(\beta_1^*, \beta_2^*) - [\frac{\delta^2 SSE}{\delta \beta_1 \delta \beta_2}(\beta_1^*, \beta_2^*)]^2$$

$$= 4 \sum_{i=1}^{n} f_1(\mathbf{x}_i)^2 \sum_{i=1}^{n} f_2(\mathbf{x}_i)^2 - 4(\sum_{i=1}^{n} f_1(\mathbf{x}_i) \cdot f_2(\mathbf{x}_i))^2$$

$$> 0$$

we can conclude that (β_1^*, β_2^*) is indeed a minimization point.

The operator proposed in this section is also non-degenerative regarding training error, since we are optimizing parameters over a set that includes $(\beta_1 = 1, \beta_2 = 0)$ and $(\beta_1 = 0, \beta_2 = 1)$.

4   Experimental Results

The experiments reported in this section were performed to evaluate the role of geometric semantic crossover on a large set of datasets with distinct properties. The first experiment (reported in Sect. 4.1) was designed to show that, different from traditional crossover operators, semantic geometric operators have nothing to do with a macro mutation [1], as they guarantee their offspring will be semantically intermediate to its parents, and they also outperform strictly mutation-based methods. The second experiment, showed in Sect. 4.2, compares variations of convex semantic crossover operators using different distances and optimized coefficients.

For all experiments, we used 20 datasets with distinct properties. Eight of these dataset are synthetic and were recommended in [7], the others being real-world datasets. For each real-world dataset, we did a 5-fold cross-validation with 10 replications, making 50 replications. For the synthetic ones (except *keijzer-6* and *keijzer-7*), we generated 5 samples and, for each sample, applied the algorithms 10 times, again making 50 replications. For *keijzer-6* and *keijzer-7*, the test set is fixed, so we simply replicated the executions 50 times.

For all methods, a preliminary parameter study was performed, and we defined the population size equal to 1,000 individuals, evolved for 2,000 generations to ensure convergence. The operator set included basic arithmetic operations: addition, subtraction, multiplication and protected division. The terminal set included the variables of the problem and constant values in the interval $[-1, 1]$. The tournament size was defined as 10. Finally, both probabilities of crossover and mutation were defined as 0.5.

It is important to point out that, in all Semantic Genetic Programming (SGP) versions, the semantic mutation operator used was implemented as in [5]. This is because this mutation operator presented better results in preliminary tests than the mutation operator proposed in [8]. We believe this difference is due to the fact that the semantic impact of the latter is still unbounded, which is not true for the mutation operator used in this work. The mutation step required by the mutation operator was defined as 10 % of the standard deviation of the training data. For each algorithm, the following variations of semantic crossover were tested:

- **SGXE:** Euclidean-based geometric semantic crossover with random crossover mask;
- **SGXM:** Manhattan-based geometric semantic crossover with random crossover mask;
- **SGXE-C:** Optimized convex Euclidean-based geometric semantic crossover operator (as in Eq. 3);
- **SGXE-L:** Optimized non-convex Euclidean-based geometric semantic crossover operator (as in Eq. 6);
- **SGP-Mut:** SGP with crossover rate equal to 0.

All statistical tests considered a confidence level of 95 %. Whenever multiple tests were necessary, a Bonferroni correction was applied to assure that the required confidence level was maintained.

4.1   Measuring the Impact of Geometric Semantic Crossover

This section compares SGP-Mut and SGXM, the most recent geometric semantic crossover operator proposed in literature. The reason for this comparison is

Table 1. Median RMSEs (and IQR) obtained after 2,000 generations for each dataset, considering 50 replications.

Dataset	SGP-Mut		SGXM	
	Median	IQR	Median	IQR
airfoil	2.28	0.16	2.65	0.91
bioavailability	33.09	3.39	30.63	4.48
concrete	5.61	0.80	4.92	0.50
cpu	37.22	9.82	30.09	12.42
energyCooling	1.34	0.18	1.19	0.16
energyHeating	0.82	0.14	0.63	0.16
forestfires	59.87	40.37	52.55	46.07
keijzer-5	0.31	0.07	0.08	0.17
keijzer-6	0.44	0.33	0.30	0.40
keijzer-7	0.05	0.02	0.03	0.10
korns-1	207.74	47.61	106.61	138.02
korns-2	476.71	60.66	687.22	2914.61
korns-12	1.13	0.11	1.02	0.01
ppb	32.38	6.27	29.22	4.87
tower	25.46	0.71	19.13	1.01
vlad-1	0.66	2.57	2.00	5.61
vlad-4	0.40	0.15	0.21	1.77
wine-red	0.65	0.07	0.59	0.05
wine-white	0.72	0.02	0.67	0.01
yacht	1.51	0.39	1.31	0.47

that, as already stated, the (semantic) fitness landscape induced by the fitness function, the distance function and the set of solutions is quite simple: since the fitness function is actually the distance function, we have that the landscape is unimodal. This may indicate that methods based on local decisions are sufficient for achieving good solutions, and hence crossover might have similar effects to mutation.

Table 1 shows the median results of Root Mean Squared Error (RMSE) followed by the Interquartile Range (IQR) obtained by both configurations on the 20 datasets used as benchmarks. The results leave no doubts that SGXM performs better than SGP-Mut. In half of the datasets, SGXM was statistically better than SGP-Mut, while being statistically worse in only two.

These results indicate that the geometric semantic crossover operator in SGP is indeed beneficial and has a different effect from using mutation-only based methods, despite the simplicity of the fitness landscape. The poor results obtained by the mutation operator might be explained by the fact that the operator has access only to the *observed* fitness landscape (training set), and not the complete one.

4.2   Comparing Different Distance Functions and Crossover Masks

Given that geometric semantic crossover is indeed necessary, we now turn our attention to the impact of the crossover operator distance function, as well as the performance of the two operators proposed here, which work by optimizing the crossover masks. Therefore, this section compares SGXE, SGXM, SGXE-C and SGXE-L using the same 20 datasets listed in the last section.

Table 2 shows the final training error obtained by each operator on each dataset, while Table 4 shows the number of datasets where the operator positioned in the line beats the operator positioned in the column in the training set, according to Wilcoxon test. As expected, in most datasets, SGXM and SGXE-C are statistically better than SGXE. SGXE-L was consistently worse than SGXE-C (and all other operators) despite considering a larger set of possible combinations. Another interesting point is that SGXM is better than SGXE-C in 11 datasets and worse in 4 datasets, leading to the conclusion that the optimization

Table 2. Median training RMSEs (and IQR) obtained after 2,000 generations for each dataset, considering 50 replications.

Dataset	SGXE		SGXM		SGXE-C		SGXE-L	
	Median	IQR	Median	IQR	Median	IQR	Median	IQR
airfoil	1.82	0.21	1.89	0.46	1.68	0.16	1.66	0.14
bioavailability	5.12	1.17	4.67	1.44	4.77	1.43	6.12	1.01
concrete	3.19	0.14	2.61	0.15	2.96	0.15	4.26	0.17
cpu	3.49	0.61	2.03	0.38	1.80	0.38	7.65	1.57
energyCooling	0.89	0.05	0.70	0.06	0.81	0.05	1.14	0.05
energyHeating	0.42	0.03	0.36	0.06	0.37	0.04	0.65	0.08
forestfires	21.80	3.22	14.43	2.31	18.86	2.98	22.12	4.25
keijzer-5	0.03	0.01	0.03	0.01	0.03	0.01	0.05	0.01
keijzer-6	0.01	0.00	0.00	0.00	0.00	0.00	0.01	0.00
keijzer-7	0.01	0.01	0.01	0.01	0.01	0.01	0.01	0.01
korns-1	86.67	17.98	78.26	18.67	82.79	26.64	7.27	10.62
korns-2	301.20	437.22	131.13	140.62	228.16	320.80	220.05	268.34
korns-12	0.93	0.01	0.87	0.01	0.91	0.01	0.95	0.01
ppb	0.25	0.03	0.09	0.02	0.08	0.02	1.25	0.17
tower	17.45	0.29	16.58	0.57	16.93	0.29	21.20	0.84
vlad-1	0.00	0.00	0.00	0.00	0.00	0.00	0.01	0.00
vlad-4	0.03	0.00	0.02	0.00	0.03	0.00	0.05	0.00
wine-red	0.36	0.01	0.29	0.01	0.33	0.01	0.38	0.01
wine-white	0.57	0.00	0.53	0.00	0.55	0.00	0.60	0.01
yacht	0.48	0.07	0.38	0.05	0.39	0.07	0.78	0.17

Table 3. Median test RMSEs (and IQR) obtained after 2000 generations for each datasets, considering 50 replications.

Dataset	SGXE		SGXM		SGXE-C		SGXE-L	
	Median	IQR	Median	IQR	Median	IQR	Median	IQR
airfoil	2.28	0.29	2.65	0.91	2.21	0.27	2.17	0.31
bioavailability	31.06	3.85	30.63	4.48	31.49	4.48	38.06	12.00
concrete	4.82	0.44	4.92	0.50	4.68	0.50	6.19	0.91
cpu	28.95	11.95	30.09	12.42	28.04	15.74	136.51	128.12
energyCooling	1.21	0.14	1.19	0.16	1.20	0.12	1.76	1.12
energyHeating	0.59	0.09	0.63	0.16	0.55	0.07	1.51	2.06
forestfires	51.55	46.68	52.55	46.07	53.00	45.81	105.85	39.83
keijzer-5	0.07	0.19	0.08	0.17	0.09	0.20	0.19	0.35
keijzer-6	0.61	0.39	0.30	0.40	0.50	0.56	0.41	0.25
keijzer-7	0.03	0.03	0.03	0.10	0.02	0.03	8.56	15.89
korns-1	104.13	160.15	106.61	138.02	102.00	147.86	8.75	47.95
korns-2	702.62	3124.15	687.22	2914.61	930.37	3031.87	2263.33	6511.75
korns-12	1.03	0.01	1.02	0.01	1.04	0.01	1.02	0.01
ppb	29.21	5.49	29.38	5.67	30.15	4.97	44.64	54.36
tower	19.36	0.71	19.13	1.01	19.29	0.89	24.36	2.45
vlad-1	1.42	3.76	2.00	5.61	2.91	7.09	6.61	27.17
vlad-4	0.09	0.43	0.21	1.77	0.34	0.60	1.85	5.49
wine-red	0.60	0.05	0.59	0.05	0.60	0.06	0.65	0.06
wine-white	0.68	0.01	0.67	0.01	0.68	0.01	0.71	0.02
yacht	1.15	0.25	1.31	0.47	1.16	0.34	2.01	0.99

is easier when considering the Manhattan distance than when considering the Euclidean distance, despite optimized coefficients.

Tables 3 and 4 show the same information as the two previous tables, now considering test error. From these tables, we observe that SGXE-L is by far the worst operator: it lost in 15 datasets and won in only 2. We attribute these results to overfitting: for instance, on the *cpu* dataset, the training error achieved by SGXE-L was 7.65, while the test error was 105.85. As expected, the removal of the convex property increased risk of overfitting, as we eliminated any bounds on the test error.

On the test set, SGXE, SGXM and SGXE-C achieved similar results. These results indicate that both SGXE-C and SGXM may lead to overfitting, since the good results obtained in the training set were not replicated in the test set. In fact, this situation is even worse for SGXM, which obtained results slightly worse than SGXE-C in the test set despite winning in the majority of datasets when considering training error.

Therefore, we conclude that the distance function and the use of optimized coefficients reduce training error drastically, but does not have the same impact when considering test error. We also observe that the flexibility gained by using

Table 4. Number of datasets where the operator presented in the line was statistically better than the operator presented in the column according to a Wilcoxon test considering training and test error.

	Training Error				Test Error			
	SGXE	SGXM	SGXE-C	SGXE-L	SGXE	SGXM	SGXE-C	SGXE-L
SGXE	0	0	0	14	0	2	1	15
SGXM	19	0	11	17	2	0	2	15
SGXE-C	18	4	0	16	1	6	0	15
SGXE-L	3	3	1	0	2	2	2	0

linear combinations instead of convex combinations in the crossover operator was not worth the loss of the convex property, which exhibited interesting results regarding control of overfitting. Based on these results, we suggest the use of SGXE, as it achieved similar results of more complex operators.

5   Conclusions

This work performed an extensive evaluation of the effects of the use of different distance functions when defining the semantic distance between two symbolic regression functions. It also proposed two new versions of the traditional operators by optimizing the coefficients involved in the convex and linear combinations of solutions.

Experimental results indicated that the use of a Euclidean or Manhattan distance function for semantic geometric crossover has little impact on test error, even when using our proposed versions with optimized coefficients. The use of linear combinations instead of convex combinations led to poor results, mainly attributed to the lack of any property regarding generalization. For SGP practitioners, we suggest the use of SGXE, as it achieved similar results to those obtained by more complex operators.

References

1. Angeline, P.J.: Subtree crossover: building block engine or macromutation. Genet. Program. **97**, 9–17 (1997)
2. Arnaldo, I., Krawiec, K., O'Reilly, U.M.: Multiple regression genetic programming. In: Proceedings of the 2014 Conference on Genetic and Evolutionary Computation, pp. 879–886. ACM (2014)
3. Beadle, L., Johnson, C.G.: Semantically driven crossover in genetic programming. In: IEEE World Congress on Computational Intelligence, pp. 111–116 (2008)
4. Beadle, L., Johnson, C.G.: Semantic analysis of program initialisation in genetic programming. Genet. Program. Evolvable Mach. **10**(3), 307–337 (2009)
5. Castelli, M., Silva, S., Vanneschi, L.: A C++ framework for geometric semantic genetic programming. Genetic Programming and Evolvable Machines (2014). http://dx.doi.org/10.1007/s10710-014-9218-0

6. Krawiec, K., Lichocki, P.: Approximating geometric crossover in semantic space. In: Proceedings of the 11th Annual conference on Genetic and evolutionary computation, pp. 987–994. ACM (2009)

7. McDermott, J., White, D.R., Luke, S., Manzoni, L., Castelli, M., Vanneschi, L., Jaskowski, W., Krawiec, K., Harper, R., De Jong, K., et al.: Genetic programming needs better benchmarks. In: Proceedings of the Fourteenth International Genetic and Evolutionary Computation Conference, pp. 791–798. ACM (2012)

8. Moraglio, A., Krawiec, K., Johnson, C.G.: Geometric semantic genetic programming. In: Coello Coello, C.A., Cutello, V., Deb, K., Forrest, S., Nicosia, G., Pavone, M. (eds.) PPSN XII. LNCS, vol. 7491, pp. 21–31. Springer, Heidelberg (2012)

9. Uy, N.Q., Hoai, N.X., O'Neill, M., McKay, R.I., Galván-López, E.: Semantically-based crossover in genetic programming: application to real-valued symbolic regression. Genet. Program. Evolvable Mach. **12**(2), 91–119 (2011)

10. Uy, N.Q., O'Neill, M., Hoai, N.X., McKay, B., Galván-López, E.: Semantic similarity based crossover in GP: the case for real-valued function regression. In: Collet, P., Monmarcé, N., Legrand, P., Schoenauer, M., Lutton, E. (eds.) Artifical Evolution. LNCS, vol. 5975, pp. 170–181. Springer, Heidelberg (2010)

11. Vanneschi, L., Castelli, M., Manzoni, L., Silva, S.: A new implementation of geometric semantic GP and its application to problems in pharmacokinetics. In: Krawiec, K., Moraglio, A., Hu, T., Şima Etaner-Uyar, A. (eds.) Genetic Programming. LNCS, vol. 7831, pp. 205–216. Springer, Heidelberg (2013)

12. Vanneschi, L., Castelli, M., Silva, S.: A survey of semantic methods in genetic programming. Genet. Program. Evolvable Mach. **15**(2), 195–214 (2014)

13. Vanneschi, L., Castelli, M., Manzoni, L., Silva, S.: A new implementation of geometric semantic GP and its application to problems in pharmacokinetics. In: Krawiec, K., Moraglio, A., Hu, T., Şima Etaner-Uyar, A., Hu, B. (eds.) Genetic Programming. LNCS, vol. 7831. Springer, Heidelberg (2014)

Learning Text Patterns Using
Separate-and-Conquer Genetic Programming

Alberto Bartoli, Andrea De Lorenzo, Eric Medvet$^{(\boxtimes)}$, and Fabiano Tarlao

DIA, Università degli Studi di Trieste, Trieste, Italy
{bartoli.alberto,andrea.delorenzo,emedvet}@units.it,
fabiano.tarlao@phd.units.it

Abstract. The problem of extracting knowledge from large volumes of unstructured textual information has become increasingly important. We consider the problem of extracting text slices that adhere to a syntactic pattern and propose an approach capable of generating the desired pattern automatically, from a few annotated examples. Our approach is based on Genetic Programming and generates extraction patterns in the form of regular expressions that may be input to existing engines without any post-processing. Key feature of our proposal is its ability of discovering automatically whether the extraction task may be solved by a single pattern, or rather a set of multiple patterns is required. We obtain this property by means of a separate-and-conquer strategy: once a candidate pattern provides adequate performance on a subset of the examples, the pattern is inserted into the set of final solutions and the evolutionary search continues on a smaller set of examples including only those not yet solved adequately. Our proposal outperforms an earlier state-of-the-art approach on three challenging datasets.

Keywords: Regular expressions · Multiple pattern · Programming by example · Text extraction

1 Introduction

The problem of extracting knowledge relevant for an end user from large volumes of unstructured textual information has become increasingly important over the recent years. This problem has many different facets and widely differing complexity levels, ranging from counting the number of occurrences of a certain word to extracting entities (e.g., persons and places) and semantics relations between them (e.g., lives-in). In this work, we are concerned with the *extraction* of text slices that adhere to a *syntactic pattern*. In particular, we investigate the feasibility of a framework where the pattern is to be generated automatically from a few *examples* of the desired extraction behavior provided by an end user.

A crucial difficulty involved in actually implementing a framework of this sort consists in generating a pattern that does not overfit the examples while at the same time providing high precision and recall on the full dataset to be processed. This difficulty is magnified when the syntactic features of the examples

© Springer International Publishing Switzerland 2015
P. Machado et al. (Eds.): EuroGP 2015, LNCS 9025, pp. 16–27, 2015.
DOI: 10.1007/978-3-319-16501-1_2

are hardly captured adequately by a single pattern. For example, dates may be expressed in a myriad of different formats and learning a single pattern capable of expressing all these formats may be very difficult. Similarly, one might want to extract, e.g., dates and IP addresses, or URLs and Twitter hashtags. The learning machinery should be able to realize automatically, based on the expressiveness of the specific pattern formalism used, how many patterns are needed and then it should generate each of these patterns with an appropriate trade-off between specificity and generality.

In this paper we describe a system based on Genetic Programming that is capable of supporting a framework of this sort, by generating *automatically* text extractor patterns in the form of *regular expressions*. The user provides a text file containing a few text slices to be extracted, which have to be annotated, and the system automatically generates a set of regular expressions, where each element is specialized for a partition of the examples: processing a text stream with all these regular expressions will implement the desired extraction behavior. From an implementation point of view, our system actually generates a single regular expression composed of several regular expressions glued together by an OR operator. This choice allows using the generated expression with existing regex processing engines, e.g., those commonly used in Java or JavaScript, without any post-processing.

A key feature of our proposal is that the system does not need any hint from the user regarding the number of different patterns required for modelling the provided examples. Depending on the specific extraction task, thus, the system automatically discovers whether a single pattern suffices or a set of different patterns is required and, in this case, of which cardinality. We obtain this property by implementing a *separate-and-conquer* approach [10]. Once a candidate pattern provides adequate performance on a subset of the examples, the pattern is inserted into the set of final solutions and the evolutionary search continues on a smaller set of examples including only those not yet solved adequately. Of course, turning this idea into practice is difficult for a number of reasons, including the identification of suitable criteria for identifying the "adequate" level of performance.

We assess our proposal on several extraction tasks of practical complexity: dates expressed in many different formats to be extracted from bills enacted by the US Congress; URLs, Twitter citations, Twitter hashtags to be extracted from a corpus of Twitter posts; and IP addresses and dates expressed in different formats to be extracted from email headers. Our approach exhibits very good performance and significantly improves over a baseline constituted by an earlier proposal, the improvement being threefold: the generated patterns (i) exhibit better extraction precision and recall on unseen examples, (ii) are simpler, and (iii) are obtained with lower computational effort.

This work builds upon an earlier proposal for generating regular expressions automatically from annotated examples and counterexamples [4]. The cited work greatly improved over the existing state of the art for automatic generation of regular expressions for text processing [1,2,6,7,12,15,19]. However, the cited proposal was designed for generating a *single* pattern capable of describing all the examples: as such, it is unable to effectively cope with scenarios requiring multiple patterns. The present work extends the cited proposal from a conceptual and from a practical point of view: the ability of discovering that multiple different patterns are required may greatly extend the scope of technologies for automatic pattern generation from examples.

Our separate-and-conquer approach bears several similarities to earlier approaches for rule induction, that aimed at synthesizing decision trees for solving classification problems [3,18]: partial solutions with adequate performance on some partition are found with an evolutionary search; the data sample is recursively partitioned according to performance-related heuristics; and, the final solution is constructed by assembling the partial solutions. In fact, our approach might be modelled as a single design point amongst those that were analyzed by hyper-heuristic evolutionary search in the design space of rule induction for classification [18]. While such a point of view may be useful, it is important to remark that text classification and text extraction are quite different problems: the former may allow partitioning input units in two classes, depending on whether they contain relevant slices (e.g., [5,13,16]); the latter also requires identifying the boundaries of the slice—or *slices*—to be extracted.

The learning of text extractor patterns might be seen as a form of *programming by examples*, where a program in a given programming language is to be synthesized based on a set of input-output pairs. Notable results in this area have been obtained for problems of string manipulation solved by means of languages much richer than regular expressions [11,14,17]. The cited works differ from this proposal since (i) they output programs rather than regular expressions, (ii) they are tailored to fully specified problems, i.e., they do not need to worry about overfitting the data, and (iii) they exploit active learning, i.e., they assume an oracle exists which can mark extraction errors in order to improve the learning process.

2   Problem Statement

A text *pattern* p is a predicate defined over strings: we say that a string s *matches* the pattern p if and only if $p(s)$ is true.

A *slice* x_s of a string s is a substring of s. A slice is identified by its starting and final indexes in the associated string. For ease of presentation, we will denote slices by their starting index and content, and we will specify the associated string implicitly. For instance, $bana_0$, na_2, an_3 and na_4 are all slices of the string **banana**. Slices of the same string are totally ordered by their starting index— e.g., na_2 precedes an_3. We say that x_s is a *superslice* of a slice x'_s (and x'_s is a *subslice* of x_s) if (i) x'_s is shorter than x_s, (ii) the starting index of x_s is smaller than or equal to the starting index of x'_s, and (iii) the final index of x_s is greater than or equal to the final index of x'_s—e.g., $bana_0$ is a superslice of na_2. We say that a slice x_s *overlaps* a slice x'_s of the same string if the intervals of the indexes delimited by their starting and ending indexes have a non empty intersection—e.g., $bana_0$ overlaps $nana_2$.

An *extraction* of a set P of patterns in a string s is a slice x_s which meets the following conditions: (i) x_s matches a pattern in P, (ii) for each superslice x'_s of x_s, x'_s does not match any pattern in P, and (ii) for each other slice x'_s which overlaps x_s, either x_s precedes x'_s or x'_s does not match any pattern in P. We denote with $e(s, P)$ the set of all the extractions of P in s. For instance, let $s =$ I_said_I_wrote_a_ShortPaper and P a set of only one pattern which describes (informally) "a word starting with a capital letter", then $e(s, P) = \{\text{I}_0, \text{I}_7, \text{ShortPaper}_{17}\}$. Note that the slices Short_{17} and Paper_{22} do not belong to $e(s, P)$, despite matching a pattern in P, as they do not meet the condition ii and iii above, respectively.

Finally, an *example* is a pair (s, X) where s is a string and X is a set of non-overlapping slices of s.

Based on the above definitions, the problem of learning a set of patterns from examples is defined as follows: given two sets of examples (E, E'), generate a set P of patterns using *only* E so that (i) the F-measure of P on E' is maximized and (ii) the complexity of P is minimized. The F-measure of P on E' is the harmonic mean of precision $\text{Prec}(P, E')$ and recall $\text{Rec}(P, E')$, which are defined as follows:

$$\text{Prec}(P, E') := \frac{\sum_{(s,X)\in E'} |e(s, P) \cap X|}{\sum_{(s,X)\in E'} |e(s, P)|} \tag{1}$$

$$\text{Rec}(P, E') := \frac{\sum_{(s,X)\in E'} |e(s, P) \cap X|}{\sum_{(s,X)\in E'} |X|} \tag{2}$$

The complexity of the set P of patterns depends on the formalism which is used to represents a pattern. In this work, we represent patterns by means of *regular expressions* and assume that the complexity of a regular expression is fully captured by its length. Hence, the complexity of P is given by $\ell(P) := \sum_{p\in P} \ell(p)$, where $\ell(p)$ is the length of the regular expression represented by p.

3   Our Approach

We propose the use of Genetic Programming for solving the problem of learning a set of patterns—in the form of a set of regular expressions—from examples. An individual of the evolutionary search is a tree which represents a regular expression and we use common GP operators (crossover and mutation) in order to generate new individuals.

We learn a set of patterns according to a *separate-and-conquer* strategy, i.e., an iterative procedure in which, at each iteration, we learn a single pattern and then remove from the set of examples those which are "solved" by the learned pattern, repeated until no more examples remain "unsolved". At the end, the learned set of patterns is composed of the patterns learned at each iteration.

We describe the single evolutionary search (i.e., one iteration) in the next section and our separate-and-conquer strategy in Sect. 3.2.

3.1 Pattern Evolutionary Search

A pattern evolutionary search takes as input a *training set* $\mathcal{T}$ and outputs a single pattern p. The training set is composed of *annotated strings*, i.e., of tuples (s, X_d, X_u), where X_d and X_u are sets of non-overlapping slices of string s (i.e., no slice in X_d overlaps any slice in X_u). Slices in X_d are desired extractions of $\{p\}$ in s, whereas slices in X_u are undesired extractions of $\{p\}$ in s.

Our pattern evolutionary search is built upon the approaches proposed in [4,5,8], which we extend in three key aspects: (i) different fitness definitions (we use three objectives rather than two objectives); (ii) different fitness comparison criteria (we use a hierarchy between the fitness indexes rather than a Pareto-ranking); and, (iii) a mechanism for enforcing diversity among individuals.

An individual is a tree which represents a regular expression, i.e., a candidate pattern. The set of terminal nodes is composed of: (i) predefined ranges a-z, A-Z and 0-9; (ii) character classes \w and \d; (iii) digits 0, ..., 9; (iv) partial ranges obtained from the slices in $\bigcup_{(s, X_d, X_u) \in \mathcal{T}} X_d$ according to the procedure described in [5]—a partial range being the largest interval of characters occurring in a set of strings (e.g., a-c and l-n are two partial ranges obtained from {cabin, male}), see the cited paper for full details); (v) other special characters such as \., :, @, and so on. The set of function nodes is composed of: (i) the concatenator •∙; (ii) the character class [•] and negated character class [^•]; (iii) the possessive quantifiers •*+, •++, •?+ and •{•,•}+; (iv) the non-capturing group (?:•). A tree represents a string by means of a depth-first post order visit in which the • symbols in a non-terminal node are replaced by the string representations of its children.

The initialization of the population of n_{pop} individuals is based on the slices in $\bigcup_{(s, X_d, X_u) \in \mathcal{T}} X_d$, as follows (similarly to [8]). For each slice $x_s \in \bigcup_{(s, X_d, X_u) \in \mathcal{T}} X_d$, two individuals are built: one whose string representation is equal to the content of x_s where each digit is replaced by \d and each other alphabetic character is replaced by \w; another individual whose string representation is the same as the former and where consecutive occurrences of \d (or \w) are replaced by \d++ (or \w++). For instance, the individuals \d-\w\w\w-\d\d and \d-\w++-\d++ are built from the slice whose content is 7-Feb-2011. If the number of individuals generated from $\mathcal{T}$ is greater than n_{pop}, exceeding individuals are removed randomly; otherwise, if it is lower than n_{pop}, missing individuals are generated randomly with a Ramped half-and-half method. Whenever an individual is generated whose string representation is not a valid regular expression, it is discarded and a new one is generated.

Each individual is a candidate pattern p and is associated, upon creation, with a *fitness* tuple $f(p) := (\mathrm{Prec}(p, \mathcal{T}), \mathrm{Acc}(p, \mathcal{T}), \ell(p))$—the first and second components are based on two operators $\sqcap$ and $\ominus$ defined over sets of slices as follows. Let X_1, X_2 be two sets of slices of the same string s. We define two operations between such sets. $X_1 \ominus X_2$ is the set of all the slices of s which (i) are a subslice of or equal to at least one slice in X_1, (ii) do not overlap any slice in X_2, and (iii) have not a superslice which meets the two first conditions. $X_1 \sqcap X_2$ is the set of all the slices of s which (i) are a subslice of or equal to at least one slice in X_1, (ii) are a subslice of or equal to at least one slice in X_2, and (iii) have not a superslice which meets the two first conditions. For instance, let $X_1 = \{I_0, I_7, \mathtt{ShortPaper}_{17}\}$ and $X_2 = \{I_0, \mathtt{Paper}_{22}\}$, then $X_1 \ominus X_2 = \{I_7, \mathtt{Short}_{17}\}$, $X_1 \sqcap X_2 = \{I_0, \mathtt{Paper}_{22}\}$.

The first fitness component is the precision on the annotated strings:

$$\mathrm{Prec}(p, \mathcal{T}) := \frac{\sum_{(s, X_d, X_u) \in \mathcal{T}} |e(s, \{p\}) \cap X_d|}{\sum_{(s, X_d, X_u) \in \mathcal{T}} |e(s, \{p\}) \sqcap (X_d \cup X_u)|} \qquad (3)$$

The second component $\mathrm{Acc}(p, \mathcal{T})$ is the average of the True Positive Character Rate (TPCR) and True Negative Character Rate (TNCR):

$$\mathrm{TPCR}(p, \mathcal{T}) := \frac{\sum_{(s, X_d, X_u) \in \mathcal{T}} \|e(s, \{p\}) \sqcap X_d\|}{\sum_{(s, X_d, X_u) \in \mathcal{T}} \|X_d\|} \qquad (4)$$

$$\mathrm{TNCR}(p, \mathcal{T}) := \frac{\sum_{(s, X_d, X_u) \in \mathcal{T}} \|(s \ominus e(s, \{p\}) \sqcap X_u\|}{\sum_{(s, X_d, X_u) \in \mathcal{T}} \|X_u\|} \qquad (5)$$

where $\|X\|$ is the sum of the length of all the slices in X.

We compare individuals using a lexicographical order on their fitness tuples (also called *multi-layered fitness* [9]): between two individuals, the one with the greatest Prec is considered the best; in case they have the same Prec, the one with the greatest Acc is considered the best; in case, finally, they have the same Prec and Acc, the one with the lowest ℓ is considered the best. Figure 1 shows an example of the fitness of two individuals on an annotated string and shows which one is the best, according to the comparison criterion here defined.

$$s = \mathtt{10th\_lap\_lasted\_from\_7:02:11\_to\_11:10:13\_of\_02-03-79}$$
$$X_d = \{7\!:\!02\!:\!11_{21}, 11\!:\!10\!:\!13_{32}\}$$
$$X_u = \{\mathtt{10th\_lap\_lasted\_from}_0, \mathtt{\_of\_}_{28}, \mathtt{\_of\_}_{40}\}$$
$$p_1 = \mathtt{\backslash d\{2,2\}+.\backslash d\{2,2\}+.\backslash d\{2,4\}+}$$
$$p_2 = \mathtt{(?:[0-9]++-?+)++)}$$
$$e(s, \{p_1\}) = \{11\!:\!10\!:\!13_{32}, 02\text{-}03\text{-}79_{44}\}$$
$$e(s, \{p_2\}) = \{10_0, 7_{21}, 02_{23}, 11_{26}, 11_{32}, 10_{35}, 13_{38}, 02\text{-}03\text{-}79_{44}\}$$
$$f(p_1) = \left(1, 0.77 = \frac{1}{2}\left(\frac{8}{15} + \frac{29}{29}\right), 26\right)$$
$$f(p_2) = \left(0, 0.83 = \frac{1}{2}\left(\frac{11}{15} + \frac{27}{29}\right), 17\right)$$

Fig. 1. Example of the fitness of two individuals p_1 and p_2 on an annotated string (s, X_d, X_u): according to our fitness comparison criterion, p_1 is better than p_2.

The population P is iteratively evolved as follows. At each iteration (generation), $0.1n_{\mathrm{pop}}$ new individuals are generated at random with a Ramped half-and-half method, $0.1n_{\mathrm{pop}}$ new individuals are generated by mutation and $0.8n_{\mathrm{pop}}$ are generated by crossover. Mutation and crossover are the classic genetic operators applied to one or two individuals selected in P with a tournament selection: $n_{\text{tour}} = 7$ individuals are chosen at random in P and the best one is selected. Whenever an individual is generated whose string representation is the same as an existing individual, the former is discarded and a new one is generated—i.e., we enforce *diversity* among phenotypes. Among the resulting $2n_{\text{pop}}$ individuals, the best n_{pop} are chosen to form the new population. The procedure is stopped when either n_{gen} iterations have been executed or the fitness tuple of the best individual has remained unchanged for more than n_{stop} consecutive iterations.

The resulting pattern p is the one corresponding to the best individual at the end of the evolutionary search.

3.2   Separate-and-Conquer Strategy

We generate a set of patterns according to a separate-and-conquer strategy [10]. We execute an iterative procedure in which, at each iteration, we execute the pattern evolutionary search described in the previous section and then remove from the training set the slices correctly extracted by the set of patterns generated so far.

In order to avoid overfitting (i.e., in order to avoid generating a set P which performs well on E yet poorly on E'), we partition the set E of examples of the problem instance in two sets E_t and E_v. The partitioning is made just once, before executing the actual iterative procedure, and is made randomly so that the number of the slices in the training and validation sets are roughly the same, i.e., $\sum_{(s,X)\in E_t} |X| \approx \sum_{(s,X)\in E_v} |X|$. The training set E_t will be used by several independent executions of the iterative procedure, whereas E_v will be used (together with E_t) to assess the pattern sets obtained as outcomes of those executions and select just one pattern set as the final solution.

In detail the iterative procedure is as follows. Initially, let the set of patterns P be empty and let $\mathcal{T}$ include all the examples in the training set E_t: for each $(s, X) \in E_t$, a triplet $(s, X, \{s\} \ominus X)$ is added to $\mathcal{T}$ (i.e., $X_d := X$ and $X_u := \{s\} \ominus X$). Then, the following sequence of steps is repeated.

1. Apply an evolutionary search on $\mathcal{T}$ and obtain p.
2. If $\text{Prec}(p, \mathcal{T}) = 1$, then set $P := P \cup \{p\}$, otherwise terminate.
3. For each $(s, X_d, X_u) \in \mathcal{T}$, set $X_d := X_d \setminus e(s, \{p\})$;
4. If $\bigcup_{(s,X_d,X_u)\in\mathcal{T}} X_d$ is empty, then terminate.

In other words, at each iteration we aim at obtaining a pattern p with perfect precision (step 2). This pattern will thus extract only slices in X_d (i.e., slices which are indeed to be extracted) but it might miss some other slices. The next iterations will target the slices which are missed by p (step 3).

We insist on generating a pattern with perfect precision at each iteration because, if a pattern extracted something wrong, no other pattern could correct that error. As a consequence, we chose to use a multi-layered fitness during the evolutionary search, where the most prominent objective is exactly to maximize $\text{Prec}(p, \mathcal{T})$.

We execute the above procedure n_{job} independent times by varying the random seed (the starting set $\mathcal{T}$ remains the same) and we obtain n_{job} possibly different sets of patterns. At the end, we choose the one with the highest F-measure on $E = E_t \cup E_v$.

4   Experimental Evaluation

We assessed our approach on three challenging datasets. *Bills* is composed of 600 examples where each string is a portion of a bill enacted by the US Congress and the slices corresponds to dates represented in several formats. *Tweets* is composed of 50000 examples where each string is the text of a tweet (Twitter post) and the slices corresponds to URLs, Twitter citations and hashtags. Finally, *Headers* is composed of 101 examples where each string is the header of an email message and the slices corresponds to IP addresses and dates represented in several formats. We built the Bills dataset by crawling the web site of the US Congress and then applying a set of regular expressions to extract dates: we made this dataset available online[1] for easing comparative analysis. The Tweets and Headers datasets are derived from those used in [4]: the strings are the same but the slices are different. Table 1 shows 10 slices for each dataset, as a sample of the different formats involved.

Table 1. A sample of the slices in the the three datasets.

Bills	Tweets	Headers
18.12.2013	@joshua_seaton	10.236.182.42
2007/01/09	#annoyed	Thu, 12 Jan 2012 04:33:34 -0800
23/03/2009	http://t.co/Bw7A5sbI	93.174.66.112
14-09-2011	#Anonymous	209.85.216.53
23,July 2001	@YourAnonNews	24 Jan 2011 09:36:00 -0000
December 31, 2001	@zataz	27 Apr 2011 09:31:01.0953
2000.01.27	@_SweetDiccWilly	Mon Oct 1 13:04:58 2012
Dec 31, 1991	http://t.co/bYxJ9NAE	Mon, 01 Oct 2012 12:05:40 +0000
1997/12/31	#OpBlitzkrieg	151.76.78.168
1999-01-19	http://t.co/GrqKGECz	Mon, 1 Oct 2012 14:04:58 +0200

In order to obtain the slices from each string in a dataset, we manually built a set $P^\star$ of regular expressions which we then applied to the strings. Table 2 shows salient information about the datasets, including the number $|E \cup E'|$ of examples, the overall length $\sum_{(s,X)\in E\cup E'} \ell(s)$ of the strings, the overall number $\sum_{(s,X)\in E\cup E'} |X|$ of slices, the overall length $\sum_{(s,X)\in E\cup E'} \|X\|$ of the slices, and the number $|P^\star|$ of regular expressions used to extract the slices.

---

[1] http://regex.inginf.units.it/.

Table 2. Salient information about the datasets.

Dataset	Examples		Slices		$|P^\star|$
	Number	Length	Number	Length	
Bills	600	16510800	3085	38960	3
Tweets	50000	4344275	71621	933646	2
Headers	101	261174	1554	32022	3

We built 15 different problem instances (E, E') for each dataset, by varying the overall number $\sum_{(s,X)\in E} |X|$ of slices in E and the random seed for partitioning the available examples in E and E'—25,50,100 slices, each obtained from 5 different seeds. Then, we applied our method to each problem instance and measured the F-measure of the generated P on E'. In order to provide a baseline for the results, we also applied the method proposed in [4]—which itself significantly improved over previous works on regular expression learning from examples—to the same problem instances. Since the cited method generates a *single* pattern p, for this method we set $P := \{p\}$.

We executed the experimental evaluation with the following parameter values: $n_{\text{pop}} = 500$, $n_{\text{gen}} = 1000$, $n_{\text{stop}} = 200$ and $n_{\text{job}} = 32$. We set the same values for the baseline, with the exception of n_{stop} which is not available in that method. We found, through an exploratory experimentation, that reasonable variations in these parameter did not alter the outcome of the comparison between our method and the baseline, which is summarized in Table 3.

Table 3. Results of the experimental evaluation. Computational Effort (CE) is expressed in 10^{10} character evaluations. $|P|$ is always equal to 1 for the baseline. ΔFm is the relative improvement of F-measure (in percentage) obtained by our method with respect to the baseline.

Dataset	Num. of	Our method						Baseline					ΔFm
	slices	Prec	Rec	Fm	$\ell(P)$	$|P|$	CE	Prec	Rec	Fm	$\ell(P)$	CE	
Bills	25	0.47	0.60	0.49	56.4	3.2	2.3	0.22	0.51	0.24	26.4	2.5	104 %
	50	0.59	0.69	0.62	76.6	4.0	6.9	0.27	0.51	0.27	97.2	6.9	129 %
	100	0.68	0.81	0.73	88.6	4.6	11.3	0.41	0.52	0.39	104.6	11.6	87 %
Tweets	25	0.99	0.92	0.94	24.6	2.4	0.6	0.90	0.86	0.87	25.6	1.1	8 %
	50	0.97	0.98	0.96	22.4	2.6	1.6	0.86	0.88	0.85	27.2	2.1	13 %
	100	0.98	0.99	0.99	25.6	3.0	3.2	0.85	0.96	0.90	46.2	4.1	10 %
Headers	25	0.84	0.74	0.79	98.6	3.2	4.6	0.43	0.41	0.41	61.0	5.1	93 %
	50	0.92	0.88	0.90	116.4	3.6	7.6	0.42	0.46	0.44	54.8	7.7	104 %
	100	0.94	0.85	0.90	118.2	3.6	15.1	0.52	0.55	0.54	58.4	15.1	67 %

The most remarkable finding is the significant improvement of our method over the baseline, summarized in the rightmost column of Table 3 in the form of relative improvement of the F-measure on E' (we remark that E' are test data not available to the learning procedure). Indeed, raw results show that our method obtained a greater F-measure in each of the 45 problem instances. The improvement is sharper for the Bills and Headers datasets (0.73 vs. 0.39 and 0.90 vs. 0.54, respectively, with 100 slices in the learning examples). These datasets exhibit a broad set of formats thus the ability of our approach to automatically discover the need of different patterns, as well as of actually generating them, does make a significant difference with respect to the baseline. Furthermore, there is an improvement also for the Twitter dataset, although the baseline exhibits very high F-measure for this dataset.

Another interesting finding concerns the complexity of the generated set of patterns. Though our method may generate, for a given problem instance, a set composed of more than one pattern, whereas the baseline always generates exactly one pattern, the overall complexity $\ell(P)$ is lower with our method in 5 on 9 problem instances. The difference is more noticeable for the Bills dataset. It is also important to remark that the average number of patterns discovered and generated by our method ($|P|$ column of Table 3) is close to the number of patterns used for annotating the dataset ($|P^\star|$ column of Table 2): our separate-and-conquer strategy does succeed in appropriately splitting the problem in several subproblems which can be solved with simpler patterns.

Table 3 shows also the *computational effort* (CE) averaged across problem instances with the same number of slices. We define CE as the number of character evaluations performed by individuals while processing a problem instance— e.g., a population of 100 individuals applied to a set E including strings totaling 1000 characters for 100 generations corresponds to CE $= 10^7$. Note that this definition is independent of the specific hardware used. It can be seen that our method does not require a CE larger than the baseline. It can also be seen that for the Tweets dataset—the one for which the improvement in terms of F-measure and complexity of the solution was not remarkable—our method required a CE sensibly lower than the baseline. We think that this finding is motivated by our early termination criterion (determined by n_{stop}, see Sect. 3.1) which allows to spare some CE when no improvements are being observed during the evolutionary search.

Finally, we provide the execution time for the two methods with 25 slices, averaged over the 5 repetitions. Our method took 30 min, 3 min, and 29 min for Bills, Tweets, and Headers, respectively; the baseline took 45 min, 6 min, and 21 min, respectively. Each experiment has been executed on a machine powered with a 6 core Intel Xeon E5-2440 (2.40 GHz) equipped with 8 GB of RAM.

5   Concluding Remarks

We considered the problem of learning a set of text extractor patterns from examples. We proposed a method for generating the patterns, in the form of regular expressions, which is based on Genetic Programming. Each individual represents a valid regular expression and individuals are evolved in order to meet three objectives: maximize the extraction precision on a training set, maximize the character accuracy on the training set, and minimize the regular expression length (a proxy for its complexity). Several evolutionary searches are executed according to a iterative separate-and-conquer strategy: the examples which are "solved" at a given iteration are removed from the examples set of subsequent iterations. This strategy allows our method to automatically discover if several patterns are needed to solve a problem instance and, at the same time, to generate those patterns.

We assessed our method and compared its performance against an earlier state-of-the-art proposal. The experimental analysis, performed on several extraction tasks of practical complexity, showed that our method outperforms the baseline along three dimensions: greater extraction precision and recall on unseen examples, simpler patterns, and lower computational effort required to generate them.

References

1. Barrero, D.F., R-Moreno, M.D., Camacho, D.: Adapting searchy to extract data using evolved wrappers. Expert Syst. Appl. **39**(3), 3061–3070 (2012). http://www.sciencedirect.com/science/article/pii/S0957417411012991
2. Barrero, D.F., Camacho, D., R-Moreno, M.D.: Automatic web data extraction based on genetic algorithms and regular expressions. In: Cao, L. (ed.) Data Mining and Multi-agent Integration, pp. 143–154. Springer, Heidelberg (2009). http://www.springerlink.com/index/G1K1N12060742860.pdf
3. Barros, R.C., Basgalupp, M.P., de Carvalho, A.C., Freitas, A.A.: A hyper-heuristic evolutionary algorithm for automatically designing decision-tree algorithms. In: Proceedings of the Fourteenth International Conference on Genetic and Evolutionary Computation Conference, pp. 1237–1244. ACM (2012)
4. Bartoli, A., Davanzo, G., De Lorenzo, A., Medvet, E., Sorio, E.: Automatic synthesis of regular expressions from examples. Computer **47**(12), 72–80 (2014)
5. Bartoli, A., De Lorenzo, A., Medvet, E., Tarlao, F.: Playing regex golf with genetic programming. In: Proceedings of the 2014 Conference on Genetic and Evolutionary Computation, pp. 1063–1070. ACM (2014)
6. Brauer, F., Rieger, R., Mocan, A., Barczynski, W.: Enabling information extraction by inference of regular expressions from sample entities. In: ACM International Conference on Information and Knowledge Management, pp. 1285–1294. ACM (2011). http://dl.acm.org/citation.cfm?id=2063763
7. Cetinkaya, A.: Regular expression generation through grammatical evolution. In: International Conference on Genetic and Evolutionary Computation, GECCO, pp. 2643–2646. ACM, New York, NY, USA (2007). http://doi.acm.org/10.1145/1274000.1274089
8. De Lorenzo, A., Medvet, E., Bartoli, A.: Automatic string replace by examples. In: Proceeding of the fifteenth annual conference on Genetic and evolutionary computation, pp. 1253–1260. ACM (2013)
9. Eggermont, J., Kok, J.N., Kosters, W.A.: Genetic programming for data classification: partitioning the search space. In: Proceedings of the 2004 ACM Symposium on Applied Computing, pp. 1001–1005. ACM (2004)

10. Fürnkranz, J.: Separate-and-conquer rule learning. Artif. Intell. Rev. **13**(1), 3–54 (1999)
11. Gulwani, S.: Automating string processing in spreadsheets using input-output examples. In: Proceedings of the 38th Annual ACM SIGPLAN-SIGACT Symposium on Principles of Programming Languages, POPL 2011, pp. 317–330. ACM, New York, NY, USA (2011). http://doi.acm.org/10.1145/1926385.1926423
12. Kinber, E.: Learning regular expressions from representative examples and membership queries. Grammatical Inference: Theoretical Results and Applications, pp. 94–108 (2010). http://www.springerlink.com/index/4T83103160M9PQ74.pdf
13. Lang, K.J., Pearlmutter, B.A., Price, R.A.: Results of the Abbadingo one DFA learning competition and a new evidence-driven state merging algorithm. In: Honavar, V., Slutzki, G., Slutzki, G. (eds.) Grammatical Inference. LNCS, vol. 1433, pp. 1–12. Springer, Heidelberg (1998). http://link.springer.com/chapter/10.1007/BFb0054059
14. Le, V., Gulwani, S.: Flashextract: A framework for data extraction by examples. In: Proceedings of the 35th ACM SIGPLAN Conference on Programming Language Design and Implementation, p. 55. ACM (2014)
15. Li, Y., Krishnamurthy, R., Raghavan, S., Vaithyanathan, S., Arbor, A.: Regular Expression Learning for Information Extraction. Computational Linguistics, pp. 21–30 (October, 2008). http://portal.acm.org/citation.cfm?doid=1613715.1613719
16. Lucas, S.M., Reynolds, T.J.: Learning deterministic finite automata with a smart state labeling evolutionary algorithm. IEEE Trans. Pattern Anal. Mach. Intell. **27**(7), 1063–1074 (2005). http://ieeexplore.ieee.org/xpls/abs_all.jsp?arnumber=1432740
17. Menon, A., Tamuz, O., Gulwani, S., Lampson, B., Kalai, A.: A machine learning framework for programming by example. In: Proceedings of the 30th International Conference on Machine Learning (ICML-13), pp. 187–95 (2013). http://machinelearning.wustl.edu/mlpapers/papers/ICML2013_menon13
18. Pappa, G.L., Freitas, A.A.: Evolving rule induction algorithms with multi-objective grammar-based genetic programming. Knowl. Inf. Syst. **19**(3), 283–309 (2009)
19. Wu, T., Pottenger, W.: A semi-supervised active learning algorithm for information extraction from textual data. J. Am. Soc. Inf. Sci. Technol. **56**(3), 258–271 (2005)

Improving Geometric Semantic Genetic Programming with Safe Tree Initialisation

Grant Dick[✉]

Department of Information Science, University of Otago,
Dunedin, New Zealand
grant.dick@otago.ac.nz

Abstract. Researchers in genetic programming (GP) are increasingly looking to semantic methods to increase the efficacy of search. Semantic methods aim to increase the likelihood that a structural change made in an individual will be correlated with a change in behaviour. Recent work has promoted the use of geometric semantic methods, where offspring are generated within a bounded interval of the parents' behavioural space. Extensions of this approach use random trees wrapped in logistic functions to parameterise the blending of parents. This paper identifies limitations in the logistic wrapper approach, and suggests an alternative approach based on safe initialisation using interval arithmetic to produce offspring. The proposed method demonstrates greater search performance than using a logistic wrapper approach, while maintaining an ability to produce offspring that exhibit good generalisation capabilities.

Keywords: Genetic programming · Semantic methods · Interval arithmetic · Safe initialisation · Symbolic regression

1 Introduction

One of the key properties of traditional forms of genetic programming (GP) is the concept of *closure*: the ability of a given element in a program to be moved into different parts of the program, or indeed into another program, and preserve its functional correctness [6]. Closure allowed GP to use straightforward implementations of its search operators, as they did not need to acknowledge the semantic properties of the subtrees upon which they acted. In their canonical form, crossover and mutation pick, swap, and replace subtrees at random. By focusing solely on syntactical correctness, rather than preserving semantic properties, these operators are not designed to produce changes in offspring that are highly correlated with their parents' behaviour.

Recently, there has been increased interest in the use of *semantic* methods to improve the basic search characteristics of GP. One particular group of semantic methods, *geometric semantic GP* (GSGP), make use of specially-designed operators that are guaranteed to produce offspring that lie within the interval bounded by their parents' behaviour [8]. GSGP has been extended by other researchers,

© Springer International Publishing Switzerland 2015
P. Machado et al. (Eds.): EuroGP 2015, LNCS 9025, pp. 28–40, 2015.
DOI: 10.1007/978-3-319-16501-1_3

specifically by the use of logistic functions that wrap around mutant subtrees, or parameterise the blending of parents during crossover [4,9].

Previous work in GP has explored the use of interval arithmetic to predict the range of outputs a parse tree will produce on unseen data [5]. Intervals are established around the input features of the problem, and used to compute the intervals of subtrees as they are executed. Typically, these intervals were used within evaluation to identify individuals that were likely to produce invalid results on unseen data, and penalise them accordingly.

This paper examines the use of logistic wrappers in GSGP. Specifically, we highlight limitations encountered through the use of logistic functions to regulate the shape of offspring produced in GSGP. We then propose a safe initialisation procedure that uses interval arithmetic to promote the generation of valid offspring. Through a number of experiments, we show that this approach is able to preserve the effective search properties of GSGP while also promoting good generalisation to previously unseen instances.

The remainder of this paper is structured as follows: Sect. 2 examines relevant items of previous work; Sect. 3 explores the use of logistic wrappers in GSGP, and highlights their limitations; Sect. 4 introduces the interval arithmetic-based safe tree initialisation method; Sect. 5 compares the safe initialisation variant of GSGP to other GP methods; finally, Sect. 6 concludes the paper with a brief discussion and suggestions for future work.

2 Semantic Methods in GP

The traditional approach to search in GP is through subtree crossover and mutation. In these operations, subtrees are picked within individuals, and either replaced with new subtrees, or swapped with subtrees from other individuals that have be picked via a similar process. This ability of general replacement of subtrees in standard GP is brought about through the *closure* requirement that the outputs of GP subtrees must be functionally interchangeable, and execute correctly in all contexts [6]. However, the closure requirement says nothing about the *behavioural* requirements of subtrees, so while a subtree may produce valid outputs regardless of its position within a tree, the context in which it resides may lead it to produce rather specialised outputs. This leads to the undesirable situation where search performance is compromised when subtrees with differing *semantic* properties are interchanged.

Recent work has seen an increased interest in the use of explicit semantic methods in GP, where search operators are guided in ways that attempt to preserve behavioural aspects of parents within offspring. There exist a number of ways that this can be done: a recent review of such methods classifies them into diversity driven, indirect and direct methods [10]. Insufficient space precludes an in-depth review of this work, so readers are pointed to this review for more details on semantic methods.

2.1 Geometric Semantic Methods in GP

This paper focuses on the direct approach of semantic methods proposed by Moraglio et al. This approach, typically referred to as *geometric semantic GP*, preserves the behavioural aspects of parents by generating offspring that are guaranteed to exist within the interval bounded by the parents' behaviours [8]. GSGP operates by encapsulating entire individuals and embedding them directly in offspring, rather than through direct manipulation of their structure as in standard GP. This ensures that semantic properties are preserved in offspring. GSGP was shown to be effective in several problem domains, although in this paper we will limit discussion to symbolic regression.[1]

Crossover in GSGP can take on two forms, labelled *SGXE* and *SGXM*. SGXE takes two parent solutions T_1 and T_2 and produces an offspring according to:

$$O_c = pT_1 + (1-p)T_2, \tag{1}$$

where p is a random real constant in $[0, 1]$. SGXM differs slightly from SGXE in that p is replaced by a random function T_R that has a co-domain $[0, 1]$. Constraining p and T_R to this interval ensures that offspring resulting from crossover lie within the region of behaviour bounded by the selected parents.

Mutation in GSGP was defined by Moraglio et al. as a single operator, *SGMR*. The operator generates two mutant trees, R_1 and R_2 and combines them with the parent solution T_1 using a fixed mutation step (ms) parameter:

$$O_m = T_1 + ms\,(R_1 - R_2)\,. \tag{2}$$

Selection of a suitable ms ensures that offspring lie within a small "ball" of behaviour centred on the parent's behaviour.

The encapsulation approach taken by GSGP allows it to guide the evolutionary search process towards changes that are *behaviourally* effective. However, the encapsulation of entire individuals means that program sizes can grow at an exponential rate. Moraglio et al. initially proposed the use of program simplification to reduce the complexity of individuals to make execution of GSGP more feasible. Subsequent research has incorporated memoization and caching of execution results to make GSGP more computationally efficient [7,9].

3 Logistic Wrappers in GSGP

Recent work has proposed extensions to the basic GSGP model in an attempt to make it more computationally efficient, and more amenable to general application. One such example is GSGP-C++, which caches the results of tree execution to improve performance [4,9]. Caching is made possible by the encapsulating properties of the GSGP operators. Additionally, GSGP-C++ uses a modified variant of the SGXM operator, and defines a new mutation operator.

---

[1] Given that over a third of GP research investigates or uses symbolic regression directly, this is not considered a serious limitation [11].

Both crossover and mutation make extensive use of logistic functions to map the execution of subtrees into [0, 1] space. In the case of crossover, the logistic function serves to ensure that offspring are created within the region of behaviour bounded by their parents. In the case of mutation, each of the two mutant subtrees are wrapped in logistic function calls to constrain their outputs to [0, 1]. This has the effect of promoting small, incremental changes as a result of mutation: in each mutation, the change in the program's behaviour is limited to the interval $[-ms, ms]$.

The logistic function takes two arguments: the input vector X, and a GP parse tree T_R. It then maps the output of T_R from processing X into a bounded [0, 1] space:

$$\text{logis}(T_R, X) = \frac{1}{1 + e^{-T_R(X)}}. \tag{3}$$

Wrapping the tree T_R in the logistic function has a couple of interesting properties: large positive outputs from T_R are mapped close to 1, while large negative outputs are mapped to 0. Raw outputs close to 0 are mapped around 0.5.

3.1   Limitations of the Logistic Wrapper Approach

At first glance, the use of logistic functions as wrapper around the randomly generated trees in GSGP seems like a elegant solution. However, further analysis suggests a number of factors that may present problems with their use:

1. The effective range of the logistic function is in the interval $[-5, 5]$. At the extremes of this interval, the logistic function quickly asymptotes to either 0 or 1, meaning that crossover behaves much like a replication operator of one of the parents, and mutation behaves more like adding a constant value to the output of the tree. Therefore, the random trees generated in crossover and mutation are only effective if they regularly produce results in $[-5, 5]$.
2. It is implicitly assumed that the random trees that are created produce results that are centred around zero.
3. If the distribution of results produced by the underlying trees are centred around zero, then the crossover operator has a peculiar behaviour that, for the region below zero, the behaviour of one parent is emphasised, while results greater than zero will place greater emphasis on the second parent. This is in contrast to the constant scale approach (SGXE), where the offspring will be a uniform blend of both parents throughout the entire input space.
4. If the underlying tree produces a non-monotonic function of the input space (e.g., it incorporates a trigonometric function), then the resulting offspring will also be a non-monotonic blend of the parents, and resemble a high-order polynomial. This may run the risk of over-fitting to the underlying data source and compromise generalisation performance.

The nature of these limitations is visualised in Fig. 1. The visualisations only examine crossover, but similar behaviour would be observed in mutation as well.

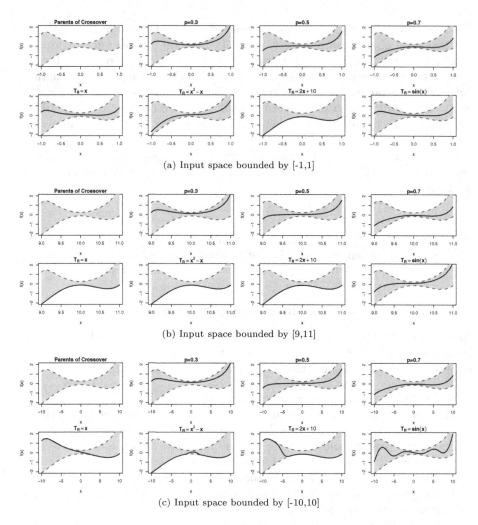

Fig. 1. Impact of a logistic wrapper over a random function approach in geometric semantic crossover.

In each plot, the shaded area represents the interval between two parents (indicated by dashed lines) in which GSGP will create offspring (indicated by solid lines). The top row of each block of figures represents the behaviour of the SGXE operator under several configurations of p, whereas the second row of each block demonstrates the behaviour of SGXM when using the logistic wrapper and several possible functions. In the first block of figures, the input space is centred around zero, and has a small overall interval of $[-1, 1]$. Here, the logistic wrappers behave well and create balanced offspring within the interval of the parents. The exception to this is the random tree that produces the function $2x + 10$, where the tree produces large outputs, and so the logistic wrapper maps consistently close to one. The result of this is that the offspring is essentially a clone of one of its parents. In the second block of figures, the width of the interval is the same, but the interval is now centred on 10. Here, the use of the logistic wrapper now produces results that bias heavily towards one parent, effectively negating crossover. The final block of images centres the input space interval on zero again, but expands the bounds to $[-10, 10]$. Here, the logistic wrapper biases towards different parents on each side of zero for most functions, but in the case of the tree that produces the function $\sin(x)$, the non-monotonic nature of the function becomes apparent, creating a somewhat chaotic looking individual from the two source parents. It is interesting to note that in all the examples shown in Fig. 1, the SGXE operator did not demonstrate any sensitivity to the nature of the input space interval. In practical terms, this would likely result in easier application to previously unexplored problems, as there would not be the same requirement to consider the nature of the input space, the likely range of the outputs, and so on.

The visualisations presented in Fig. 1 are designed to clearly demonstrate the theoretical limitations of using logistic wrappers. To test the practical consequences of these limitations, we generated 100 random trees for three benchmark regression data sets from previous work. For each of these trees, we executed them on the instances in their respective data sets (outlined in Sect. 5), and then applied the logistic wrapper. The resulting histograms are presented in Fig. 2. As can be seen, the first data set, Bioavailability, presents the majority of its outputs at 0.5, which suggests a raw output before logistic wrapping of zero. This is explained by the anomalous properties of the data set, in which approximately half the fields in the data set contain no information (they are completely filled with zeros), so any tree that uses solely these fields as terminals will always return zero. The second most common value in the bioavailability outputs was 0.731, which corresponds to a raw output of one. For the other two data sets — Concrete and Yacht Hydrodynamics — the vast majority of outputs are in the 0 and 1 bins; in these scenarios, the effect of crossover would be lost and behaviourally it would resemble reproduction, while for mutation it would amount to adding a constant to the existing individual.

The conclusion that can be drawn from this exploration is that using logistic wrappers to constrain random tree outputs to $[0, 1]$ may introduce unwanted consequences that may hinder the search process in GSGP.

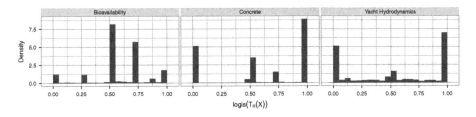

Fig. 2. Distributions of outputs emitted by randomly-generated trees, and subsequently transformed by a logistic function wrapper, on three different problems.

4   Interval Arithmetic for Safe Tree Initialisation

The previous section revealed some challenges with using a logistic wrapper over randomly-generated trees to weight parent components in GSGP. However, without the logistic wrapper to constrain the space of tree outputs, the default approach to GSGP proposed by Moraglio et al. shares a significant issue with standard GP: if the instances supplied to GP for training do not sufficiently capture the underlying distribution of the input features, then the evolved GP model may produce uncharacteristically wild outputs on some unseen inputs (e.g., division by zero). The approach proposed by Castelli et al. alleviates this problem, but at the expense of search performance. What is needed is a way to incorporate domain knowledge about the anticipated domains of inputs, and use this to guide tree construction and search.

Interval arithmetic has been previously used in GP to evolve solutions more likely to behave sanely on unseen data [5]. Intervals are simply the set of real numbers that an input can take within a given upper and lower bound. Using these bounds, we can define the intervals that result from operations such as addition and division, and define the intervals upon which a given operation (e.g., division) is valid. We can use basic knowledge of the intervals of a problem's input features to guide the search process in GP. For example, interval arithmetic can be used during tree execution: intervals defined for each input are used to construct the interval for each subtree (based upon its rooted operation) as it is executed. When invalid intervals are encountered during execution, the tree can be flagged as not valid over the entire input space, and penalised accordingly (e.g., assigned a very poor fitness).

Algorithm 1. `BuildTree`: the safe initialisation tree generation algorithm.

Require: *depth*: the current tree depth; *maxdepth*: the maximum required tree depth; F: a list of functions for inner nodes; T: a list of terminals (the input features of the problem); I: the known intervals of the input features

Ensure: A tree that produces output valid within the intervals defined by its subtrees

1: **if** `PickTerminal`(*depth*, *maxdepth*, $|F|$, $|T|$) **then**
2:    $op \leftarrow$ random element from T
3:    **return** $\{root = op, left = \emptyset, right = \emptyset, interval = I[op]\}$
4: **else**
5:    $sub_{left} \leftarrow$ `BuildTree`(*depth* + 1, *maxdepth*, F, T, I)
6:    $sub_{right} \leftarrow$ `BuildTree`(*depth* + 1, *maxdepth*, F, T, I)
7:    $op \leftarrow$ `SelectOperation`(F, $sub_{left}[interval]$, $sub_{right}[interval]$)
8:    $int_{op} \leftarrow$ `UpdateInterval`(op, $sub_{left}[interval]$, $sub_{right}[interval]$)
9:    **return** $\{root = op, left = sub_{left}, right = sub_{right}, interval = int_{op}\}$
10: **end if**

While a similar approach to using interval arithmetic during execution could be used within the context of GSGP, this paper takes a different approach. Because the intervals do not depend upon any single instance in the data set, they can be used outside of the evaluation process, for example in tree creation. Typically, tree creation in GP follows a simple process: the operation for a given node is first selected, and then any required children subtrees are created recursively. If this sequence is reversed (i.e., the children subtrees are created before the node operation is selected), then interval arithmetic can be used to select an appropriate operation for the root node, thereby increasing the likelihood of the tree producing valid outputs over the entire input space. The process for this is defined in Algorithm 1. Note that although this paper defines the algorithm for binary operators, extensions to this algorithm for arbitrary arity functions would be fairly straightforward.

Aside from the reversed order of creating subtrees before selecting the root node's operation, the `BuildTree` method follows a standard tree creation procedure. The initialisation process described in Algorithm 1 makes use of three support routines `PickTerminal`, `SelectOperation`, and `UpdateInterval`. The operation `PickTerminal` is the same as what would be found in any standard tree generation method and decides whether the current node will be a terminal or a function. The `SelectOperation` procedure uses the intervals of the new subtrees to select an operator from our set of options. For example, if the set was $\{+, -, \times, \div\}$, and the interval for sub_{right} included zero, then we would only pick from the first three operations. Finally, `UpdateInterval` computes the overall interval of the new subtree based upon the operator selected in the previous step and the intervals of the two subtrees.

5   Results

The performance of GSGP (using the SGXE and SGMR operators described in Sect. 2.1) with the `BuildTree` safe initialisation method was compared to the GSGP-C++ framework of Castelli et al. Additionally, a 'default' implementation of GSGP without safe initialisation (but again using SXGE and SGMR) was used to provide a baseline comparison.[2] The parameters for these experiments are identical to those described in previous work and are used without modification [4]. No attempt to find ideal parameters for each method was performed, as it was assumed that the default parameters defined by Castelli et al. would constitute a suitable configuration. We chose a set of seven test problems used in previous work — refer to this work for more details on these data sets [1]. Additionally, we use the bioavailability data set from previous work, as it appears to have been used to guide the development of GSGP-C++ [2,3,9]. For the safe initialisation approach, we sourced the intervals for the input features directly from the data sets; in a real-world context defining these intervals could have also taken into account other domain knowledge, but that was not done here.

To measure the training and generalisation performance, we use the root-relative-squared-error (RRSE) defined as:

---

[2] Source code for experiments available at: https://github.com/grantdick/libgsgp.

$$RRSE(y, \hat{y}) = \sqrt{\frac{\sum_{i=1}^{|y|} (y_i - \hat{y}_i)^2}{\sum_{i=1}^{|y|} (y_i - \bar{y})^2}},\tag{4}$$

where y and $\hat{y}$ are the recorded response values of the data set and the model predictions, respectively. RRSE is essentially the root-mean-squared error standardised by the underlying deviance of the response variable, and makes for easy comparison of the relevance of a model's performance — an RRSE of one is functionally equivalent to a model that predicts a fixed value (the mean of the data) for all inputs, while an RRSE of zero is presented by a perfect model. Questions should normally be asked of a model-building process that routinely creates models with an RRSE greater than one. Unless otherwise stated, all RRSE results presented here are for *test results* on previously unseen data. To determine training and testing splits, we used a repeated 10-fold cross-validation approach. A total of 10 cross-validation processes were performed, for a total of 100 runs for each combination of GP configuration and problem instance.

The evolution of training and testing fitness over time for the three data sets explored in Sect. 3.1 is shown in Fig. 3; with respect to the logistic wrapper, recall that the histogram associated with the Yacht Hydrodynamics data set (Fig. 2) exhibited the most extreme distribution, with almost all values presented being either zero or one. The Concrete data set was next, with most values being either zero or one, but with a smaller number presenting a value of 0.5. Finally, the bioavailability data set presented the least number of extreme values. The results presented in Figs. 2 and 3 correlate well for the logistic wrapper: on the

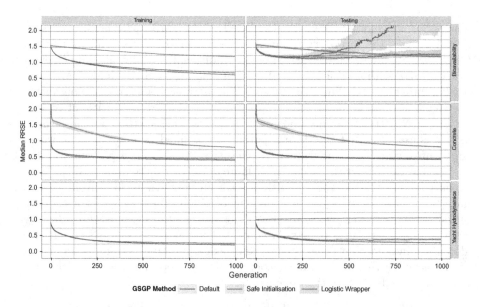

Fig. 3. Median training and testing RRSE performance of the three GSGP methods. The coloured bands around each series is the 95 % confidence interval of the median.

Table 1. Proportion of runs with a test RRSE below a given threshold. Ideally, RRSE values should be in [0,1], so larger values indicate catastrophic model fitting.

Problem	GSGP Method	Proportion of runs with RRSE				
		≤ 1	≤ 10	≤ 100	≤ 1000	≤ 10000
Bioavailability	Default	0.03	0.62	0.77	0.82	0.87
	Safe Initialisation	0.15	0.92	0.99	0.99	1.00
	Logistic Wrapper	0.00	1.00	1.00	1.00	1.00
Concrete	Default	0.99	1.00	1.00	1.00	1.00
	Safe Initialisation	1.00	1.00	1.00	1.00	1.00
	Logistic Wrapper	0.98	1.00	1.00	1.00	1.00
Yacht Hydrodynamics	Default	0.79	0.86	0.87	0.88	0.88
	Safe Initialisation	1.00	1.00	1.00	1.00	1.00
	Logistic Wrapper	0.00	1.00	1.00	1.00	1.00

bioavailability problem, a large number of values around 0.5 will be generated, which means that crossover and mutation with a logistic wrapper will actually produce novel offspring, and hence the search is more effective. Conversely, because the distribution of values in the Yacht Hydrodynamics problem are highly biased towards zero and one, the search operators (in particular, crossover) will be ineffective, and so search will stagnate. The two GSGP approaches that do not use logistic wrapping demonstrate similar search characteristics to each other, and demonstrate a faster rate of evolution than the logistic wrapper approach. However, the GSGP approach without safe initialisation exhibits a high degree of catastrophic over-fitting on the bioavailability problem. The safe initialisation approach to GSGP does not exhibit such behaviour.

Table 1 characterises the degree to which the three GSGP methods exhibit over-fitting and poor generalisation. It can be seen that, for the three problems examined, GSGP with safe initialisation demonstrates the greatest ability to find solutions with a testing RRSE less than one. Conversely, the GSGP approach using logistic wrappers appears to be adversely affected: while it does not demonstrate any great degree of catastrophic over-fitting, it also appears to have difficulty evolving solutions that present a meaningful RRSE below one. Finally, the default implementation of GSGP demonstrates a high degree of catastrophic over-fitting on two of the three problems, suggesting that safe initialisation is effective at controlling this phenomena and promoting good generalisation performance.

Finally, we compare the generalisation performance of the three GSGP methods on the seven benchmark data sets from previous work. For a point of reference, a standard GP approach is also included. The results of this analysis are shown in Table 2. In addition to this analysis, a Kruskal-Wallis rank sum test, with post hoc test for differences, was performed. The post-hoc tests suggest that only three of the interactions do not show a significant difference: the safe

Table 2. Quantiles of the testing RRSE values of the three GSGP methods on known benchmark problems. Results for standard GP are included for reference.

Problem	GSGP Method	Testing RRSE Quartiles:				
		0 %	25 %	50 %	75 %	100 %
Boston Housing	*Standard GP*	5.31E-01	6.51E-01	7.10E-01	8.08E-01	2.95E+13
	Default	3.94E-01	5.04E-01	5.74E-01	7.32E-01	3.17E+00
	Safe Initialisation	3.22E-01	4.73E-01	5.18E-01	5.99E-01	3.46E+00
	Logistic Wrapper	3.91E-01	6.08E-01	6.78E-01	7.28E-01	9.95E-01
Concrete	*Standard GP*	4.71E-01	6.06E-01	7.63E-01	8.62E-01	2.32E+01
	Default	3.62E-01	4.25E-01	4.58E-01	4.91E-01	1.78E+00
	Safe Initialisation	3.76E-01	4.41E-01	4.72E-01	5.14E-01	5.89E-01
	Logistic Wrapper	7.04E-01	8.10E-01	8.49E-01	9.03E-01	1.12E+00
Dow Chemical	*Standard GP*	7.47E-01	9.34E-01	9.87E-01	1.00E+00	1.52E+01
	Default	6.89E-01	7.87E-01	8.66E-01	9.76E-01	9.49E+00
	Safe Initialisation	7.08E-01	7.75E-01	8.04E-01	8.42E-01	1.30E+00
	Logistic Wrapper	4.84E-01	5.75E-01	2.08E+00	1.02E+01	3.69E+01
Energy Efficiency	*Standard GP*	8.20E-02	1.66E-01	2.21E-01	3.18E-01	4.58E-01
	Default	1.32E-01	2.54E-01	2.73E-01	2.95E-01	3.63E-01
	Safe Initialisation	2.07E-01	2.59E-01	2.80E-01	3.02E-01	3.53E-01
	Logistic Wrapper	1.72E-01	2.56E-01	2.95E-01	3.22E-01	5.05E-01
Parkinsons	*Standard GP*	9.49E-01	9.78E-01	9.98E-01	1.12E+00	3.43E+01
	Default	9.34E-01	9.51E-01	9.61E-01	9.83E-01	7.97E+15
	Safe Initialisation	9.26E-01	9.39E-01	9.47E-01	9.55E-01	1.00E+00
	Logistic Wrapper	9.90E-01	1.02E+00	1.03E+00	1.04E+00	1.07E+00
Wine Quality (Red)	*Standard GP*	8.25E-01	9.02E-01	9.73E-01	1.13E+00	9.47E+10
	Default	7.06E-01	8.00E-01	8.27E-01	8.64E-01	2.82E+11
	Safe Initialisation	7.25E-01	7.87E-01	8.07E-01	8.42E-01	9.32E-01
	Logistic Wrapper	7.67E-01	8.22E-01	8.53E-01	8.83E-01	9.61E-01
Yacht Hydrodynamics	*Standard GP*	8.49E-02	1.54E-01	2.30E-01	3.49E-01	8.98E+12
	Default	1.67E-01	3.14E-01	4.07E-01	6.75E-01	3.56E+12
	Safe Initialisation	1.87E-01	2.51E-01	3.09E-01	3.49E-01	6.71E-01
	Logistic Wrapper	1.04E+00	1.07E+00	1.08E+00	1.10E+00	1.75E+00

initialisation and default GSGP methods on the Concrete data set, the safe initialisation and default GSGP methods on the Energy Efficiency data set, and the safe initialisation and logistic wrapper GSGP methods also on the Energy Efficiency data set. All other comparisons were considered significant to a 95 % confidence level. Interestingly, for two of the problems, Energy Efficiency and Yacht Hydrodynamics, none of the GSGP approaches could match the performance of standard GP. For the other problems, GSGP with safe initialisation presented significantly lower RRSE scores.

6   Conclusion and Future Work

Recent work in genetic programming has placed significant emphasis on improving the efficacy of semantic-based methods. One branch of this work, geometric semantic genetic programming (GSGP), attempts to increase the semantic connection between parents and offspring by blending fully-encapsulated parents within the child. This paper examined an extension to the GSGP approach that wraps randomly generated trees with logistic functions before embedding them in offspring. Analysis of this approach uncovered several limitations of the use of logistic wrapper functions. To improve the search and generalisation performance, an alternative approach was developed, based around the use of interval arithmetic within tree initialisation to promote the construction of valid individuals that generalise well. When tested on a range of problems, it was shown to regularly outperform both the standard GSGP model as well as the GSGP approach using logistic wrappers.

The work presented in this paper opens up several avenues for future research. GSGP approaches typically create very large individuals that grow at an exponential rate. It would be interesting to see if the safe initialisation method has any impact on this growth rate. Likewise, it would be interesting to see how the incorporation of interval arithmetic into the tree execution process (in addition to safe initialisation) would impact on GSGP performance. The safe initialisation method is not limited to GSGP, so it would be interesting to measure its impact on standard GP. Finally, although not fully explored here, there appears to be a clear relationship between GSGP and additive models — stronger integration of the concepts of traditional machine learning approaches to GSGP may yield positive results, and so should be explored in detail.

References

1. Agapitos, A., McDermott, J., O'Neill, M., Kattan, A., Brabazon, A.: Higher order functions for kernel regression. In: Nicolau, M., Krawiec, K., Heywood, M.I., Castelli, M., García-Sánchez, P., Merelo, J.J., Rivas Santos, V.M., Sim, K. (eds.) EuroGP 2014. LNCS, vol. 8599, pp. 1–12. Springer, Heidelberg (2014)
2. Archetti, F., Lanzeni, S., Messina, E., Vanneschi, L.: Genetic programming for human oral bioavailability of drugs. In: Proceedings of the 8th Annual Conference on Genetic and Evolutionary Computation, GECCO 2006, pp. 255–262. ACM, New York, NY, USA (2006)
3. Archetti, F., Lanzeni, S., Messina, E., Vanneschi, L.: Genetic programming for computational pharmacokinetics in drug discovery and development. Genet. Program. Evolvable Mach. 8(4), 413–432 (2007)
4. Castelli, M., Silva, S., Vanneschi, L.: A C++ framework for geometric semantic genetic programming. Genet. Program. Evolvable Mach. 16(1), 73–81 (2014)
5. Keijzer, M.: Improving symbolic regression with interval arithmetic and linear scaling. In: Ryan, C., Soule, T., Keijzer, M., Tsang, E., Poli, R., Costa, E. (eds.) Genetic Programming. LNCS, vol. 2610, pp. 70–82. Springer, Heidelberg (2003)
6. Koza, J.R.: Genetic Programming: On the Programming of Computers by Means of Natural Selection. MIT Press, Cambridge, MA, USA (1992)
7. Moraglio, A.: An efficient implementation of GSGP using higher-order functions and memoization. In: Johnson, C., Krawiec, K., Moraglio, A., O'Neill, M. (eds.) Semantic Methods in Genetic Programming (13 September 2014), Workshop at Parallel Problem Solving from Nature 2014 Conference, Ljubljana, Slovenia, Springer, Heidelberg (2014)
8. Moraglio, A., Krawiec, K., Johnson, C.G.: Geometric semantic genetic programming. In: Coello, C.A.C., Cutello, V., Deb, K., Forrest, S., Nicosia, G., Pavone, M. (eds.) PPSN 2012, Part I. LNCS, vol. 7491, pp. 21–31. Springer, Heidelberg (2012)

9. Vanneschi, L., Castelli, M., Manzoni, L., Silva, S.: A new implementation of geometric semantic GP and its application to problems in pharmacokinetics. In: Krawiec, K., Moraglio, A., Hu, T., Etaner-Uyar, A.Ş., Hu, B. (eds.) EuroGP 2013. LNCS, vol. 7831, pp. 205–216. Springer, Heidelberg (2013)

10. Vanneschi, L., Castelli, M., Silva, S.: A survey of semantic methods in genetic programming. Genet. Program. Evolvable Mach. **15**(2), 195–214 (2014)

11. White, D.R., McDermott, J., Castelli, M., Manzoni, L., Goldman, B.W., Kronberger, G., Jaśkowski, W., O'Reilly, U.M., Luke, S.: Better GP benchmarks: community survey results and proposals. Genet. Program. Evolvable Mach. **14**(1), 3–29 (2013)

On the Generalization Ability of Geometric Semantic Genetic Programming

Ivo Gonçalves[1,2]([⊠]), Sara Silva[1,2,3], and Carlos M. Fonseca[1]

[1] CISUC, Department of Informatics Engineering,
University of Coimbra, 3030-290 Coimbra, Portugal
{icpg,cmfonsec}@dei.uc.pt

[2] BioISI - Biosystems and Integrative Sciences Institute, Faculty of Sciences,
University of Lisbon, Campo Grande, 1749-016 Lisbon, Portugal
sara@fc.ul.pt

[3] NOVA IMS, Universidade Nova de Lisboa, 1070-312 Lisbon, Portugal

Abstract. Geometric Semantic Genetic Programming (GSGP) is a recently proposed form of Genetic Programming (GP) that searches directly the space of the underlying semantics of the programs. The fitness landscape seen by the GSGP variation operators is unimodal with a linear slope by construction and, consequently, easy to search. Despite this advantage, the offspring produced by these operators grow very quickly. A new implementation of the same operators was proposed that computes the semantics of the offspring without having to explicitly build their syntax. This allowed GSGP to be used for the first time in real-life multidimensional datasets. GSGP presented a surprisingly good generalization ability, which was justified by some properties of the geometric semantic operators. In this paper, we show that the good generalization ability of GSGP was the result of a small implementation deviation from the original formulation of the mutation operator, and that without it the generalization results would be significantly worse. We explain the reason for this difference, and then we propose two variants of the geometric semantic mutation that deterministically and optimally adapt the mutation step. They reveal to be more efficient in learning the training data, and they also achieve a competitive generalization in only a single operator application. This provides a competitive alternative when performing semantic search, particularly since they produce small individuals and compute fast.

Keywords: Geometric semantic genetic programming · Generalization · Overfitting · Pharmacokinetics · Drug discovery

1 Introduction

Geometric Semantic Genetic Programming (GSGP) [8] is a recently proposed form of Genetic Programming (GP) [6] that searches directly the space of the underlying semantics of the programs. One of the most interesting properties of GSGP is that the fitness landscape seen by its variation operators is a cone by

© Springer International Publishing Switzerland 2015
P. Machado et al. (Eds.): EuroGP 2015, LNCS 9025, pp. 41–52, 2015.
DOI: 10.1007/978-3-319-16501-1_4

construction, and consequently easy to search. Despite this advantage, the individuals produced by these operators are always bigger than their parents. Since this growth is rather quick, GSGP ends up being hard to use in practice, specially in real-life multidimensional datasets. To counteract this, a new implementation of the same geometric semantic operators was proposed by Vanneschi et al. [10]. In this implementation, the semantics of the offspring can be determined without having to explicitly build their syntax. This allowed GSGP to be used for the first time in real-life multidimensional datasets. Results have shown that besides the expected good performance on the training data, GSGP also presented a surprisingly good generalization ability. This generalization ability was justified by the authors as a result of some properties of the geometric semantic operators. However, their implementation of the mutation operator [10] presented a small deviation from the original definition [8] that is still valid under the geometric semantic framework. This implementation is available in the free open-source GSGP C++ library [2]. In our work we study the effect of both implementations of the geometric semantic mutation on the generalization ability of GSGP. We also propose and test two new variations of the geometric semantic mutation which are able to provide an optimal mutation step adaptation.

The paper is organized as follows. Section 2 contextualizes GSGP. Section 3 describes the experimental setup. Section 4 presents and discusses the effect of both implementations of the mutation operator on the generalization ability of GSGP. Section 5 proposes and discusses the results of the two new geometric semantic mutation operators, and Sect. 6 concludes.

2   Geometric Semantic Genetic Programming

Moraglio et al. [8] recently proposed a new GP formulation called Geometric Semantic Genetic Programming (GSGP). GSGP derives its name from the fact that it is formulated under a geometric framework [7] and from the fact that it operates directly in the space of the underlying semantics of the individuals. In this context, semantics is defined as the outputs of an individual over a set of data instances. Perhaps the most interesting property of GSGP is that the fitness landscape seen by its variation operators is always unimodal with a linear slope (cone landscape) by construction. This implies that there are no local optima, i.e., with the exception of the global optimum, every point in the search space has at least one neighbor with better fitness and that neighbor is reachable through the application of the variation operators. The immediate consequence of this type of landscape is that it is easy to search. A drawback of GSGP is that its operators always produce offspring bigger that their parents. Since our work is on regression problems, the GSGP operators presented here are for real-value semantics. For proofs and further details the reader is referred to [8].

Definition 1 (Geometric Semantic Crossover). *Given two parent functions* $T_1, T_2 : \mathbb{R}^n \to \mathbb{R}$, *the geometric semantic crossover returns the real function* $T_{XO} = (T_1 \cdot T_R) + ((1 - T_R) \cdot T_2)$, *where* T_R *is a random real function whose output values range in the interval* $[0, 1]$.

From the crossover definition it follows that every offspring is bigger than its parents combined. This leads to exponential individual growth.

Definition 2 (Geometric Semantic Mutation). *Given a parent function* $T : \mathbb{R}^n \rightarrow \mathbb{R}$, *the geometric semantic mutation with mutation step* ms *returns the real function* $T_M = T + ms \cdot (T_{R1} - T_{R2})$, *where* T_{R1} *and* T_{R2} *are random real functions.*

For the mutation operator, the individual growth produced is linear. The continuous individual growth produced by both operators renders GSGP hard to use in practice, specially in real-life multidimensional datasets. Vanneschi et al. [10] tackled this issue by providing a different implementation of these operators. In this implementation, the semantics of the offspring can be determined without having to explicitly build their syntax. Consequently, Vanneschi et al. [10] were able to use GSGP for the first time in real-life multidimensional datasets. They reported competitive performance both on training and testing data. The arguments presented for the good performance on testing data will be presented and discussed in Sect. 4.2.

However, the implementation of the mutation operator of Vanneschi et al. [10] had a small deviation from the original definition. Their implementation imposed that the random subtrees generated (T_{R1} and T_{R2}), always had a logistic function as their root node. This implies that the output of each random subtree ranges in the interval $[0, 1]$ and that, consequently, the output resulting from subtracting these random subtrees ranges in the interval $[-1, 1]$. As the mutation operator applies a mutation step, the final output added to each parent always ranges in the interval $[-ms, ms]$. Looking back at the original definition of the geometric semantic mutation [8], there is no defined range for the outputs of the random subtrees. It should be noted that this small implementation deviation is still valid under the geometric semantic framework. This deviation was not explicit in their work but was confirmed upon contact, and it is also the implementation made available in the GSGP library [2]. For clarification purposes, we will refer to the original mutation definition as Unbounded Mutation (UM) and to the alternative mutation implementation as Bounded Mutation (BM). In the end, BM applies a structural bound on the perturbation applied to the parent. This bound holds independently of the data (training or testing). We explore the effects of using a structural bound in Sect. 4.

3   Experimental Setup

To provide a fair comparison between unbounded and bounded mutation, our experimental setup is similar to the one of Vanneschi et al. [10]. The experimental parameters are provided in Table 1. The mutation step for GSGP is set to 1 as this showed better results in the preliminary testing than the value of 0.001 used by Vanneschi et al. [10]. Experiments are run for 500 generations because that is where the statistical comparisons were made in the mentioned work. Standard GP and Semantic Stochastic Hill Climber (SSHC) [8] are used as baselines for comparison. GSGP without crossover is also tested (GSGP NC). As this work studies the effects of unbounded and bounded mutations (UM and BM respectively), each method is tested with both mutations. Therefore, the variants tested are: GSGP UM and BM; GSGP NC UM and BM and SSHC UM and BM. All claims of statistical significance are based on Mann-Whitney U tests, with Bonferroni correction, and considering a significance level of $\alpha = 0.05$. For each dataset 30 different random partitions are used. Each variant uses the same 30 partitions. Experiments are conducted on the same two multidimensional symbolic regression real-life datasets used by Vanneschi et al. [10]. These datasets are the Bioavailability (hereafter Bio) and the Plasma Protein Binding (hereafter PPB). They have, respectively, 359 instances and 241 features, and 131 instances and 626 features. For a detailed description of these datasets the reader is referred to Archetti et al. [1] and Vanneschi et al. [10].

Table 1. GSGP and Standard GP parameters used in the experiments

Parameter	Value
Runs	30
Generations	500
Population size	100
Training - Testing division	70 % - 30 %
Fitness	Root Mean Squared Error
GSGP crossover	SGXM [8], probability 0.5
GSGP mutation	SGMR [8], probability 0.5
Standard GP crossover	Standard subtree crossover, probability 0.9
Standard GP mutation	Standard subtree mutation, probability 0.1
Tree initialization	Ramped Half-and-Half, maximum depth 6
Function set	+, -, *, and /, protected as in [9]
Terminal set	Input variables, no constants
Parent selection	Tournament of size 4
Elitism	Best individual always survives
Maximum tree depth	None

4   Experimental Study

All the evolution plots presented in the next sections are based on the median over 30 runs of the training and testing error of the best individuals in the training data. The median was preferred over the mean since it is more robust to outliers. Section 4.1 presents the results and Sect. 4.2 discusses the generalization ability.

4.1   Results

Figure 1 presents the training and testing error evolution plots in both datasets. This figure also shows the adaptive mutation step variants (SSHC AUM, SSHC DAUM, SSHC ABM and SSHC DABM) that will be presented and discussed in Sect. 5.

Starting with the comparisons against Standard GP, it was confirmed that GSGP BM generalizes better in both datasets (p-values: Bio 1.794×10^{-6} and

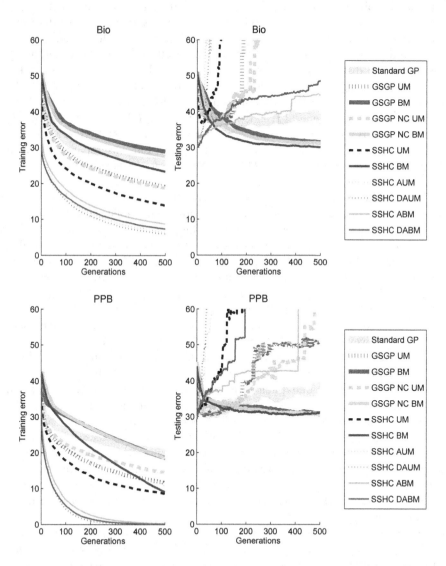

Fig. 1. Bio (top) and PPB (bottom) training and testing error evolution plots

PPB 5.121×10^{-4}). These same conclusions were already presented by Vanneschi et al. [10]. In terms of training error, Standard GP is superior to GSGP BM in the Bio dataset (p-value 5.434×10^{-5}) and no statistically significant differences were found in the PPB dataset. On the other hand, Standard GP is superior to GSGP UM in terms of testing error in the Bio dataset (p-value 9.273×10^{-4}), while no statistically significant differences were found in the PPB dataset. GSGP UM is, however, superior to Standard GP in terms of training error in both datasets (p-values: Bio 2.872×10^{-11} and PPB 2.872×10^{-11}).

It can be observed in evolution plots that Standard GP and GSGP UM overfit the training data, while GSGP BM generalizes well. In these datasets, Standard GP is known to overfit (e.g., [4]) and GSGP BM has been recently shown, and also confirmed here, to generalize well [10]. On the other hand, there is a clear distinction between GSGP BM and UM, as the latter quickly overfits the training data. It generalizes even worse than Standard GP in the Bio dataset. This same distinction between BM and UM occurs with GSGP NC and with SSHC. As a general trend, the BM variants (GSGP BM, GSGP NC BM and SSHC BM) generalize well, while the UM variants (GSGP UM, GSGP NC UM and SSHC UM) overfit the training data. This discrepancy between the generalization ability of the UM and BM variants is discussed in the next section.

Vanneschi et al. [10] mentioned that GSGP requires a relatively high mutation probability in order to explore the search space more efficiently. Indeed, our results show only small differences between using the crossover operator (GSGP UM and BM) or not using it at all (GSGP NC UM and BM). Statistically, there are no differences in terms of generalization, in any comparison with the same mutation operator. In terms of training error the results are not consistent: GSGP NC BM is significantly better than GSGP BM on the Bio dataset (p-value 3.955×10^{-5}); GSGP UM is significantly better than GSGP NC UM on the PPB dataset (p-value 4.734×10^{-11}); and no other statistically significant differences were found. However, a possible inefficiency of the crossover operator should be expected. This operator can only produce an offspring which improves over both parents when the target semantics are between (even if partially) the semantics of the parents. Without an explicit semantic diversity control of the population and a mate selection procedure that takes the target semantics into account, the crossover operator may be inefficient. This inefficiency may also increase with larger semantic spaces, i.e., as the number of data instances increases. From these experiments, it can be concluded that the crossover operator can be skipped altogether since it does not significantly and consistently improve the search outcome (in testing or training error). It also presents the disadvantage of exponentially increasing the size of the individuals, as opposed to the linear increase with the mutation operator.

On a final note, the evolution plots also show that the SSHC variants consistently learn faster than the GSGP and GSGP NC variants with the same mutation operator. This should be expected as the semantic space has no local optima and consequently the search can be focused around the best individual in the population. This leads to a faster decrease in the training error and, in the case of SSHC BM, to a faster generalization to unseen data.

4.2 Discussion on Generalization Ability

From the results presented in Sect. 4.1, it is clear that what differentiates the several methods in terms of generalization is the usage of a bounded or unbounded mutation (BM and UM respectively). BM variants generalize well, while UM variants overfit the training data.

The GSGP implementation of Vanneschi et al. [10] used a BM and reached the same conclusions regarding its competitive generalization. They justified this generalization ability by considering some properties of the geometric semantic operators. Particularly, they remarked that the geometric properties of those operators hold independently of the data on which the evaluation is taken place and consequently they also hold on testing data. For the crossover operator this implies that each offspring produced also stands between its parents in the testing data semantic space. Therefore, in the worst case, each offspring is not worse than the worst of its parents on testing data. The implication for the mutation operator is that the perturbation that each offspring produces is bounded, also in the testing data semantic space, by the mutation step (ms). Specifically, the semantic variation on the testing data also ranges in the interval $[-ms, ms]$. Therefore, Vanneschi et al. [10] concluded that the geometric semantic operators guarantee that a possible worsening of the testing error is bounded and consequently that these operators help control overfitting.

As seen before, the usage of a bounded or unbounded mutation was crucial in determining the generalization achieved. The BM operator was able to produce a competitive generalization by guaranteeing bounded and small perturbations on the testing data. This was crucial to generalize well. However, it is clear that perturbations that increase the testing error are always possible. It is also clear that if these perturbations were a significant majority of the applications of the operator then overfitting would be inevitable. Therefore, it can be concluded that after reaching what can be thought of as a generalization plateau (the point where it seems that no further induction can be performed with the available data), the BM operator generates about half of its perturbations in the decreasing testing error direction and the other half in the increasing testing error direction. These perturbations end up compensating each other and therefore creating the relatively smooth generalization plateau. On the other hand, the UM operator performed badly in terms of generalization. Since in this operator the perturbations produced on the testing data can be arbitrarily large, a single application of a mutation that results in overfitting (decreases training error but increases testing error) can have an arbitrarily large increase in the testing error. This results in considerable uncertainty in the testing error evolution. This effect may be more noticeable in regression problems since any data instance can have an arbitrarily large error contribution, as opposed to classification problems where normally the error is bounded for each data instance.

For these reasons, the BM operator seems more robust and should be preferred over the UM operator.

For the reasons already mentioned in Sect. 4.1, the crossover operator had little effect in the results. However, in principle, the crossover operator should be riskier in terms of generalization than the BM operator. This is because the variation in the testing data semantic space, although bounded by the testing semantics of the parents, can still be arbitrarily large. This results from the fact that the parents can be, in terms of testing data semantics, very far apart. Since there is no way of knowing if the parents are close or far apart in the testing data semantic space, the bounds (defined by the semantics of both parents), on testing data, are not useful in practice. This is another disadvantage of the crossover operator, following the exponential growth of the offspring produced and the low efficiency in terms of search.

Although the generalization achieved by the GSGP with bounded mutation is very competitive, the issue of the size of the solutions generated by these geometric semantic operators remains. As mentioned in Sect. 2, using crossover in GSGP translates into an exponential growth of the individuals. In our experimental study, individuals in GSGP reach several millions of nodes with only around 20 generations conducted. This raises the question: how can such large/complex individuals (models) achieve such competitive generalization? Some interpretations of theories such as Occam's razor and the Minimum Description Length principle state that smaller/less complex models generalize better. Consequently, and in light of this view, this result would be improbable, if not impossible. How can this be? A possible answer may lie in ensemble learning. Ensemble learning is a Machine Learning paradigm in which several models are created and combined to produce a final model. Dietterich [3] provided three reasons as to why constructing an ensemble of models may be superior to constructing a single model. The first two reasons are computational and representational. The computational reason is related to the difficulties in searching the search space, such as getting stuck in a local optima. The representational reason arises when the true target function cannot be represented by any of the hypotheses in the search space. These first two reasons are not discussed in detail as they are not relevant to GSGP, respectively because the semantic space has no local optima and because in GSGP (and in traditional GP) any hypothesis can be represented that could also be represented by an ensemble. The last reason is the one which is relevant to GSGP and to generalization in general. It is a statistical reason and it is related to the fact that several different models can have a similar or even the same training data performance. This is essentially a model selection problem. Which model should be chosen? There is no way of knowing which model will generalize better. Ensemble learning tackles this issue by combining several accurate models, which reduces the risk of the final model being overfitted. Even if some overfitted models are present in the ensemble, their negative contribution to the final model will be reduced since the final model will also include contributions from models which generalize well. It is a common result in ensemble learning to have large ensembles which achieve competitive generalization. Therefore, and in general, large/complex models (individuals) can also generalize well depending on how they are constructed.

GSGP can be seen as an ensemble learning method, since its operators always combine existing individuals independently to produce new individuals. The crossover operator combines two parents with a randomly generated individual and the mutation operator combines one parent with another randomly generated individual (the individual which results from subtracting the two random subtrees). We can think of these parents and randomly generated individuals as full models themselves. This interestingly relates back to ensemble learning, where a necessary condition for its positive outcome is that the ensemble has a mix of accurate and diverse models [5]. In GSGP we can think of the parents as the accurate models (as they have survived during the evolution) and the randomly generated individuals as providing the also needed diversity. GSGP may derive some of its competitive generalization from this. If, for instance, we consider the application of two sequential mutations, it follows that:

$$P + R1 * ms + R2 * ms$$

where P is the initial parent, R1 and R2 are the two randomly generated individuals and ms is the mutation step. Consequently, considering only the usage of the mutation operator, GSGP can be seen as a weighted sum combination of models (we can consider that the initial parent has a weight of 1).

In the end, GSGP successfully combines elements from ensemble learning (implicitly) and from the geometric semantic framework. Combining several models to incrementally produce new models has roots in ensemble learning. This allows to reduce the model selection risk by offsetting possible bad models. On the other hand, the combination of a structurally bounded mutation (BM) and a small mutation step can further reduce the issue of adding bad models by guaranteeing that their contribution will be small.

5   Adapting the Mutation Step

As discussed in the previous section, the mutation step can play a role in reducing the risk of overfitting. When it comes to learning more efficiently, the geometric semantic mutation can be improved by adapting its step. It is possible to deterministically compute the optimal mutation step for each application of the operator. The description of how this can be accomplished is presented in Sect. 5.1. Section 5.2 presents and discusses the results.

5.1   Optimal Step Adaptation

The geometric semantic mutation can be seen as a linear combination of two elements: the parent P, and the random individual RI which results from subtracting the two random subtrees. Since RI is multiplied by the mutation step ms, we want to find a mutation step such that:

$$P + RI * ms = t$$

where P and RI are semantic vectors and t is the target vector of the data. Since the parent is not influenced by any weight, we can rewrite this as:

$$\mathbf{RI} * \mathbf{ms} = (\mathbf{t} - \mathbf{P})$$

where we reach a general linear system:

$$\mathbf{Ax} = \mathbf{y}$$

The resolution of which can be performed deterministically and optimally by the application of the Moore-Penrose pseudoinverse (hereafter simply inverse). This inverse computes the mutation step which minimizes the error in the training data for each specific combination of RI, P and t. We will call this modification of the mutation operator as Adaptive Mutation (AM). As this work has studied the effects of bounded and unbounded mutations, we can divide the AM as: Adaptive Unbounded Mutation (AUM) and Adaptive Bounded Mutation (ABM).

Following a similar reasoning, another mutation operator can be devised. We can consider the possibility of adding a weight to the parent and adjusting both weights with the inverse. Let pw be the parent weight. Consequently:

$$\mathbf{P} * \mathbf{pw} + \mathbf{RI} * \mathbf{ms} = \mathbf{t}$$

This new semantic mutation operator will be called as Doubly Adaptive Mutation (DAM) and it can also be divided as: Doubly Adaptive Unbounded Mutation (DAUM) and Doubly Adaptive Bounded Mutation (DABM). The inverse method could also be used to perform a linear combination of more than two weighted individuals.

5.2   Results and Discussion

The newly devised operators were tested with the SSHC (more efficient than GSGP, see Sect. 4.1) and consequently its variants were named: SSHC AUM, SSHC DAUM, SSHC ABM and SSHC DABM. Figure 1 (in Sect. 4.1) shows the evolution of training and testing error for these adaptive variants. They reveal to be superior in terms of learning the training data when compared to the SSHC variants without adaptive mutation step (SSHC UM and SSHC BM). This was expected, since the step adaptation is optimal for each application of the operators. In terms of generalization, these variants quickly overfit. In light of the analysis made in Sect. 4.2, this quick overfitting should also be expected, as there is no structural bound coupled with a small mutation step and consequently no overfitting risk reduction. Since in these variants the weights can be arbitrarily large, the benefits of using a structural bound (SSHC ABM and SSHC DABM) are lost.

However, an interesting property can be found when looking closely at the initial generations. Figure 2 presents the testing error evolution on the first 10 generations. It shows that these variants achieve a competitive generalization in only a single application of the mutation operators. This is particularly clear in the SSHC DAUM and SSHC DABM variants. It was expected that these two variants fit the training data more easily when compared to the other two variants (SSHC AUM and SSHC ABM), since they have an extra degree of freedom (the parent weight). Further testing is needed to determine if this property holds across other datasets. If it holds, then these mutation variants become a competitive alternative when performing semantic search, particularly since they produce small individuals and compute fast. They also raise no issues in constructing/reconstructing large individuals, as opposed to what may happen with the GSGP variants.

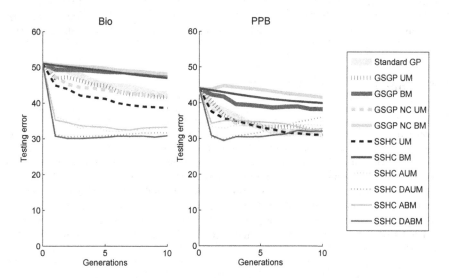

Fig. 2. Bio (left) and PPB (right) testing error evolution plots for the first 10 generations

6   Conclusions

In this work we have studied the generalization ability of Geometric Semantic Genetic Programming (GSGP), by analyzing the effects of two different implementations of the geometric semantic mutation. These implementations differ on the existence or not of a guaranteed bound on the semantic variation across both training and unseen data. Results showed that the generalization ability of GSGP differs significantly depending on whether or not this bound is used. On the tested datasets, the unbounded mutation highly overfitted the training data, while the bounded mutation achieved a competitive generalization. We have also expanded on previously reported geometric semantic arguments as to why GSGP may generalize well. Furthermore, we provided an explanation as to why such large solutions like the ones produced by GSGP can generalize competitively, by discussing how GSGP may relate with ensemble learning.

We have also proposed two new variants of the geometric semantic mutation. These new operators are able to deterministically compute the optimal mutation step for each application of the operator. They have shown to be more efficient in learning the training data, and they also achieve a competitive generalization in only a single operator application. This provides a competitive alternative when performing semantic search, particularly since they produce small individuals and compute fast.

Acknowledgments. This work was partially supported by national funds through FCT under contract UID/Multi/04046/2013 and projects PTDC/EEI-CTP/2975/2012 (MaSSGP), PTDC/DTP-FTO/1747/2012 (InteleGen) and EXPL/EMS-SIS/1954/2013 (CancerSys). The first author work is supported by FCT, Portugal, under the grant SFRH/BD/79964/2011.

References

1. Archetti, F., Lanzeni, S., Messina, E., Vanneschi, L.: Genetic programming for computational pharmacokinetics in drug discovery and development. Genet. Program. Evolvable Mach. **8**(4), 413–432 (2007)
2. Castelli, M., Silva, S., Vanneschi, L.: A C++ framework for geometric semantic genetic programming. Genet. Program. Evolvable Mach. **15**, 73–81 (2014)
3. Dietterich, T.G.: Ensemble methods in machine learning. In: Kittler, J., Roli, F. (eds.) MCS 2000. LNCS, vol. 1857, pp. 1–15. Springer, Heidelberg (2000)
4. Gonçalves, I., Silva, S.: Balancing learning and overfitting in genetic programming with interleaved sampling of training data. In: Krawiec, K., Moraglio, A., Hu, T., Etaner-Uyar, A.Ş., Hu, B. (eds.) EuroGP 2013. LNCS, vol. 7831, pp. 73–84. Springer, Heidelberg (2013)
5. Hansen, L.K., Salamon, P.: Neural network ensembles. IEEE Trans. Pattern Anal. Mach. Intell. **12**(10), 993–1001 (1990)
6. Koza, J.R.: Genetic Programming: On the Programming of Computers by Means of Natural Selection (Complex Adaptive Systems), 1st edn. The MIT Press, Cambridge (1992)
7. Moraglio, A.: Towards a geometric unification of evolutionary algorithms. Ph.D. thesis, Department of Computer Science, University of Essex, UK, November 2007
8. Moraglio, A., Krawiec, K., Johnson, C.G.: Geometric semantic genetic programming. In: Coello, C.A.C., Cutello, V., Deb, K., Forrest, S., Nicosia, G., Pavone, M. (eds.) PPSN 2012, Part I. LNCS, vol. 7491, pp. 21–31. Springer, Heidelberg (2012)
9. Poli, R., Langdon, W.B., McPhee, N.F., Koza, J.R.: A field guide to genetic programming (2008). Lulu.com
10. Vanneschi, L., Castelli, M., Manzoni, L., Silva, S.: A new implementation of geometric semantic GP and its application to problems in pharmacokinetics. In: Krawiec, K., Moraglio, A., Hu, T., Etaner-Uyar, A.Ş., Hu, B. (eds.) EuroGP 2013. LNCS, vol. 7831, pp. 205–216. Springer, Heidelberg (2013)

Automatic Derivation of Search Objectives for Test-Based Genetic Programming

Krzysztof Krawiec[✉] and Paweł Liskowski

Institute of Computing Science, Poznań University of Technology, Poznań, Poland
{krawiec,pliskowski}@cs.put.poznan.pl

Abstract. In genetic programming (GP), programs are usually evaluated by applying them to tests, and fitness function indicates only how many of them have been passed. We posit that scrutinizing the outcomes of programs' interactions with individual tests may help making program synthesis more effective. To this aim, we propose DOC, a method that autonomously derives new search objectives by clustering the outcomes of interactions between programs in the population and the tests. The derived objectives are subsequently used to drive the selection process in a single- or multiobjective fashion. An extensive experimental assessment on 15 discrete program synthesis tasks representing two domains shows that DOC significantly outperforms conventional GP and implicit fitness sharing.

Keywords: Genetic programming · Program synthesis · Test-based problems · Multiobjective evolutionary computation

1 Introduction

In genetic programming (GP), the quality of a candidate program is usually assessed by confronting it with a set of tests (fitness cases). The outcomes of program's interactions with individual tests are then aggregated by a fitness function. In discrete domains, this usually boils down to counting the number of passed tests.

Although employing a fitness function defined in this way may appear natural at first sight, there are several drawbacks of driving the search purely by the number of passed tests. Starting from not necessarily the most severe one, for n tests, fitness will take on $n + 1$ possible values, and once a search process identifies good and thus similarly fit solutions, ties become likely. Next, this quality measure is oblivious to the fact that some tests can be inherently more difficult to pass than others. But most importantly, aggregation of interaction outcomes into a single scalar implies compensation: two programs that perform very differently on particular tests may receive the same fitness and thus become indiscernible in a subsequent selection phase.

Furthermore, conventional fitness in GP is known to exhibit low fitness-distance correlation [22], i.e., it does not reflect well the number of search steps

© Springer International Publishing Switzerland 2015
P. Machado et al. (Eds.): EuroGP 2015, LNCS 9025, pp. 53–65, 2015.
DOI: 10.1007/978-3-319-16501-1_5

required to reach the optimal solution. As a result, guiding search by a fitness function defined in this way may be not particularly efficient. In other words, fitness function, despite embodying the *objective* quality of candidate solutions (considered as prospective outcomes of program synthesis process), is not necessarily the best *driver* to guide the search. Alternative *search drivers*, meant as substitutes for objective function, should be sought that correlate better with distance, possibly by reflecting other aspects of program behavior.

As we argued in [11], the habit of using scalar objective functions in domains like GP, where more detailed information on solutions' characteristic is easily available, seems particularly wasteful. The information on the outcomes of individual interactions can and should be exploited more efficiently wherever possible. In GP, search drivers could be evaluation measures that capture program's performance only on a *subset* of tests.

Various means, reviewed in Sect. 4 of this paper, have been proposed in the past to address the weaknesses of conventional fitness measure in GP. The method we propose here and describe in Sect. 3 is inspired by previous work in coevolutionary algorithms, and builds upon the approach we designed for test-based problems in [15]. In every generation, the algorithm identifies the groups of tests on which the programs in the current population behave *similarly*. Each such group gives rise to a separate *derived objective*. Typically, a few such objectives emerge from this process, and we employ them to perform selection on the current population. We propose two selection procedures that exploit the derived objectives, one of them involving the NSGA-II method [4]. In an experimental assessment reported in Sect. 5, the method performs significantly better than conventional GP and implicit fitness sharing.

2   Background

The task of automated program synthesis by means of genetic programming can be conveniently phrased as an optimization problem in which the search objective is to find a candidate solution $p^* = argmax_{p \in \mathcal{P}} f(p)$ that maximizes the objective function f, where $\mathcal{P}$ is the space of all candidate programs. In nontrivial problems, $\mathcal{P}$ is large or even infinite, and grows exponentially with the length of considered programs. Searching the entire space is therefore computationally infeasible, and one needs resort to a heuristic algorithm that is not guaranteed to find p^*. In GP, it is common to drive the search process using f as fitness function. As motivated earlier, this is not always the best approach.

A program to be evolved is typically specified by a set of tests (fitness cases). Each test is a pair $(x, y) \in T$, where x is the input fed into a program, and y is the desired outcome of applying it to x. From the machine learning perspective, T forms the training set. While in general the elements of $t \in T$ can be arbitrary objects, for the purpose of this study, we limit our interest to Boolean and integer-valued inputs and outputs.

In many problems, fitness cases do not enumerate all possible pairs of program inputs and outputs. Ideally, the synthesized program is expected to generalize beyond the training set which bears resemblance to *test-based problems*

G	t_1	t_2	t_3	t_4	t_5
a	1	1	0	1	1
b	0	1	0	1	0
c	1	0	1	1	0
d	0	1	0	0	0

a) Interaction matrix G

G	t_1	t_2	t_3	t_4	t_5
a	1	1	0	1	1
b	0	1	0	1	0
c	1	0	1	1	0
d	0	1	0	0	0

b) G after clustering

G'	t_{1+3}	t_{2+4+5}
a	0.5	1
b	0	0.66
c	1	0.33
d	0	0.33

c) Derived objectives G'

Fig. 1. Example of deriving search objectives from interaction matrix G (a) using clustering (b), resulting in the derived objectives shown in (c).

originating from the field of coevolutionary algorithms [1,3]. In test-based problems, candidate solutions interact with multiple environments – tests. Typically, the number of such environments is very large, making it infeasible to evaluate candidate solutions on all of them. Depending on problem domain, tests may take on the form of, e.g., opponent strategies (when evolving a game-playing strategy) or simulation environments (when evolving a robot controller).

In this light, it does not take long to notice that also the program synthesis task can be formulated as a test-based problem, in which passing a test requires a program to produce the desired output for a given input. In general, we will assume that an *interaction* between a program p and a test t produces a scalar outcome $g(p, t)$ that reflects the capability of the former to *pass* the latter. In this paper, we assume that interaction outcome is binary, i.e., $g : \mathcal{P} \times \mathcal{T} \rightarrow \{0, 1\}$.

A GP algorithm solving a test-based problem (program synthesis task) maintains a population of programs $P \subset \mathcal{P}$. In every generation, each program $p \in P$ interacts with every test $(x, y) \in T$, in which p is applied to x and returns an output denoted as $p(x)$. If $p(x) = y$, p is said to *solve* the test and $g(p(x), y) = 1$. If, on the other hand, $p(x) \neq y$, we set $g(p(x), y) = 0$ and say that p *fails* (x, y).

As it will become clear in the following, it is convenient to gather the outcomes of these interactions in an *interaction matrix* G. For a population of m programs and $|T| = n$, G is an $m \times n$ matrix where g_{ij} is the outcome of interaction between the ith program and jth test.

Given this test-based framework, the conventional GP fitness that rewards a program for the number of passed tests can be written as

$$f(p) = |\{t \in T : g(p, t) = 1\}|. \tag{1}$$

3   The DOC Algorithm

The proposed method of discovery of search objectives by clustering (DOC) addresses the shortcomings of conventional evaluation (cf. Sect. 1) by clustering the interaction outcomes into several *derived objectives*. Each derived objective is intended to capture a subset of 'capabilities' exhibited by the programs in the context of other individuals in population. The derived objectives replace then the conventional fitness function (Eq. 1).

Technically, DOC replaces the conventional evaluation stage of GP algorithm (cf. Sect. 2) in favor of the following steps:

1. Calculate the $m \times n$ interaction matrix G between the programs from the current population P, $|P| = m$, and the tests from T, $|T| = n$.
2. Cluster the tests. We treat every column of G, i.e., the vector of interaction outcomes of all programs from P with a test t, as a point in an m-dimensional space. A clustering algorithm of choice is applied to the n points obtained in this way. The outcome of this step is a partition $\{T_1, \ldots, T_k\}$ of the original n tests in T into k subsets/clusters, where $1 \leq k \leq n$ and $T_j \neq \emptyset$.
3. Define the derived objectives. For each cluster T_j, we average row-wise the corresponding columns in G. This results in an $m \times k$ *derived interaction matrix* G', with the elements defined as follows:

$$g'_{i,j} = \frac{1}{|T_j|} \sum_{t \in T_j} g(s_i, t) \tag{2}$$

where s_i is the program corresponding to the ith row of G, and $j = 1, \ldots, k$.

The columns of G' implicitly define the k *derived objectives* that characterize the programs in P.

The derived objectives form the basis for selecting the most promising programs from P, which subsequently give rise to the next generation of programs. The natural avenue here is to apply a multiobjective evolutionary algorithm. Following our previous work, we employ NSGA-II [4], one of the most popular method of that sort. This allows programs that feature different behaviors, reflected in the derived objectives, to coexist in population even if some of them are clearly better than others in terms of conventional fitness. However, we will show in the experimental section that such multiobjective selection may involve certain undesired side-effects, and that driving selection by certain scalar aggregate of the derived objectives can be also an interesting option.

Properties of DOC. An important property of DOC is its contextual character manifested by the fact that the outcome of evaluation of any program in P depends not only on the tests in T, but also on the other programs in P. This is the case because all programs in P together determine the result of clustering and therefore influence the derived objectives. This quite direct interaction between the programs is not a common feature of GP.

An implication of contextual evaluation is that derived objectives are *adaptive* and driven by the current state of evolving programs. The process of their discovery repeats in every generation so that they reflect the changes in behaviors of the programs in population. The derived objectives are thus *subjective* in this sense, which makes them analogue to search drivers used in two-population coevolution [15], even though the tests does not change with time here.

As clustering *partitions* the set of tests T (rather than, e.g., *selecting* some of them), none of the original tests is discarded in the transformation process. The more two tests are similar in terms of programs' performance on them, the more likely they will end up in the same cluster and contribute to the same derived objective. In the extreme case, tests that are mutually redundant (i.e., identical columns in G) are guaranteed to be included in the same derived objective.

For $k = 1$, DOC degenerates to a single-objective approach: all tests form one cluster, and G' has a single column that contains solutions' fitness as defined by Eq. 1 (albeit normalized). On the other hand, setting $k = n$ implies $G' = G$, and every derived objective being associated with a single test.

4   Related Work

There are two groups of past studies related to this work, those originating in GP and those originating in research on coevolutionary algorithms. We review these groups in the following.

In the group of **methods that originate in GP**, a prominent example of addressing the issues outlined in Sect. 1 is implicit fitness sharing (IFS) introduced by Smith *et al.* [20] and further explored for genetic programming by McKay [16,17]. IFS lets the evolution assess the difficulty of particular tests and *weighs* the rewards granted for solving them. Given a set of tests T, the IFS fitness of a program p in the context of a population P is defined as:

$$f_{IFS}(p) = \sum_{t \in T : g(p,t)=1} \frac{1}{|P(t)|} \tag{3}$$

where $P(t)$ is the subset of programs in P that solve test t, i.e., $P(t) = \{p \in P : g(p,t) = 1\}$. IFS treats tests as limited resources: programs *share* the rewards for solving particular tests, each of which can vary from $\frac{1}{|P|}$ to 1 inclusive. Higher rewards are provided for solving tests that are rarely solved by population members (small $P(t)$), while importance of tests that are easy (large $P(t)$) is diminished. The assessed difficulties of tests change as P evolves, which can help escaping local minima.

Other methods that reward solutions for having rare characteristics have been proposed as well. An example is co-solvability [10] that focuses on individual's ability to properly handle *pairs* of fitness cases, and as such can be considered a 'second-order' IFS. Such pairs are treated as elementary competences (skills) for which solutions can be awarded. Lasarczyk *et al.* [14] proposed a method for selection of fitness cases based on a concept similar to co-solvability. The method maintains a weighted graph that spans fitness cases, where the weight of an edge reflects the historical frequency of a pair of tests being solved simultaneously. Fitness cases are then selected based on a sophisticated analysis of that graph.

Last but not least, the relatively recent research on semantic GP [12] can be also seen as an attempt to provide search process with richer information of programs' behavioral characteristics. Similarly, pattern-guided GP and behavioral evaluation [13] clearly set similar goals.

In the group of studies that **originate in coevolutionary algorithms**, Pareto coevolution [6,18] was initially proposed to overcome the drawbacks of an aggregating fitness function. In Pareto coevolution, aggregation of interaction outcomes has been abandoned in favor of using each test as a separate objective. As a result, a test-based problem can be transformed into a multi-objective optimization problem. This, in turn, allows adoption of dominance relation — a candidate solution s_1 dominates a candidate solution s_2 if and only if s_1 performs at least as good as s_2 on all tests. Nevertheless, the number of such elementary objectives is often prohibitively large due to a huge number of tests present in typical test-based problems.

It was later observed that certain test-based problems feature an *internal structure* comprising groups of tests that examine the same *skill* of solutions. Based on this observation, Bucci [1] and de Jong [2] introduced *coordinate systems* that *compress* the elementary objectives into a multidimensional structure, while preserving the dominance relation between candidate solutions. Because of the inherent redundancy of tests, the number of so-called underlying objectives (dimensions) in such a coordinate system is typically lower than the number of tests. However, even with a moderately large number of tests, it is unlikely for a candidate solution to dominate any other candidate solution in the population. From such a sparse dominance relation, it is hard to elicit any information that would efficiently drive the search process. The coordinate systems introduced in the cited work do not help in this respect, as they perfectly preserve the dominance relation, and if the dominance in the original space is sparse, they need to feature very high number of dimensions. Also, the problem of their derivation is NP-hard [8].

The derived objectives constructed by DOC bear certain similarity to the underlying objectives studied in the above works. However, as shown by the example in Fig. 1, the derived objectives are not guaranteed to preserve dominance: given a pair of candidate solutions (p_1, p_2) that do not dominate each other in the original space of interaction outcomes, one of them may turn out to dominate the other in the space of resulting derived objectives. For instance, given the interaction matrix as in Fig. 1a, program c does not dominate d, however it does so in the space of derived objectives (Fig. 1c). As a result of clustering, some information about the dominance structure has been lost. This inconsistency buys us however a critical advantage: the number of resulting derived objectives is low, so that together they are able to impose an effective search gradient on the evolving population.

5   Experimental Verification

We examine the capabilities of DOC within the domain of tree-based GP. The compared algorithms implement generational evolutionary algorithm and vary only in the selection procedure. Otherwise, they share the same parameter settings, with initial population filled with the ramped half-and-half operator, subtree-replacing mutation engaged with probability 0.1 and subtree-swapping crossover engaged with probability 0.9. We run two series of experiments: one with runs lasting up to 200 generations and population size $|P| = 500$, and with runs up to 100 generations and population size $|P| = 1000$. The search process stops when the assumed number of generation elapses or an ideal program is found; the latter case is considered a success.

Table 1. Success rate (percent of successful runs) of best-of-run individuals, averaged over 30 evolutionary runs. Bold marks the best result for each benchmark

	$|P| = 500$						$|P| = 1000$					
	GP	IFS	RAND	DOC	DOC-P	DOC-D	GP	IFS	RAND	DOC	DOC-P	DOC-D
Cmp6	20	**100**	50	21	83	78	26	**97**	48	22	64	77
Cmp8	0	**56**	0	0	21	29	0	**7**	0	0	4	5
Disc1	0	0	0	7	3	**13**	0	0	0	10	10	7
Disc2	0	4	0	10	14	**37**	0	0	0	0	21	**40**
Disc3	0	0	0	18	53	**62**	0	0	0	56	71	**77**
Disc4	0	0	0	0	0	**7**	0	0	0	4	0	0
Disc5	0	0	0	0	**7**	3	0	0	0	0	**4**	**4**
Maj6	22	**100**	60	40	83	90	52	**100**	71	81	96	89
Malcev1	0	18	24	18	70	**76**	14	27	33	25	69	**93**
Malcev2	3	3	0	7	27	**30**	0	0	11	17	**32**	27
Malcev3	0	7	8	23	**83**	**83**	0	3	8	43	**93**	75
Malcev4	0	0	4	7	**10**	7	0	0	0	**25**	20	10
Malcev5	17	30	25	54	47	**57**	17	23	44	**100**	68	60
Mux6	77	**100**	83	73	**100**	**100**	90	**100**	96	**100**	**100**	**100**
Par5	0	14	14	**18**	7	12	4	6	0	**18**	3	0

Compared algorithms. The particular implementation of DOC used in this work employs X-MEANS [19], an extension of the popular k-means algorithm that autonomously adjusts k. Given an admissible range of k, X-MEANS picks the k that leads to clustering that maximizes the Bayesian Information Criterion. In this experiment, we allow X-MEANS consider $k \in [1,4]$ and employ the Euclidean metric to measure the distances between the observations (the columns of G).

We confront DOC with several control setups. The first baseline is the conventional Koza-style GP (**GP** in the following), which employs tournament of size 7 in the selection phase. The second control is implicit fitness sharing (**IFS** [17]) presented in Sect. 4, with fitness defined as in Formula 3 and also with tournament of size 7. The last control configuration, **RAND**, is a crippled variant of DOC. In that configuration, the tests, rather than being clustered based on interaction outcomes as described in Sect. 3, are partitioned into k subsets at random with k randomly drawn from the interval $[2,4]$. RAND is intended to control for the effect of multiobjective selection performed by NSGA-II (which is known to behave very differently from the tournament selection).

Benchmark problems. In its current form presented in Sect. 3, DOC can handle only binary interaction outcomes, where a program either passes a test or not. Because of that, we compare the methods on problems with discrete interaction outcomes. The first group of them are **Boolean benchmarks**, which employ instruction set $\{and, nand, or, nor\}$ and are defined as follows. For an v-bit comparator $Cmp\,v$, a program is required to return $true$ if the $\frac{v}{2}$ least significant input bits encode a number that is smaller than the number represented by the $\frac{v}{2}$ most significant bits. In case of the majority $Maj\,v$ problems, $true$ should be returned if more that half of the input variables are $true$. For the multiplexer

Mul v, the state of the addressed input should be returned (6-bit multiplexer uses two inputs to address the remaining four inputs). In the parity *Par v* problems, *true* should be returned only for an odd number of *true* inputs.

The second group of benchmarks are the **algebra problems** from Spector *et al.*'s work on evolving algebraic terms [21]. These problems dwell in a ternary domain: the admissible values of program inputs and outputs are $\{0, 1, 2\}$. The peculiarity of these problems consists of using only one binary instruction in the programming language, which defines the underlying algebra. For instance, for the a_1 algebra, the semantics of that instruction is defined as in (a) below (see [21] for the definitions of the remaining four algebras). For each of the five algebras considered here, we consider two tasks (of four discussed in [21]). In the *discriminator term* tasks (*Disc* in the following), the goal is to synthesize an expression that accepts three inputs x, y, z and is semantically equivalent to the one shown in (b) below. There are thus $3^3 = 27$ fitness cases in these benchmarks. The second tasks (*Malcev*), consists in evolving a so-called *Mal'cev term*, i.e., a ternary term that satisfies the equation (c) below. This condition specifies the desired program output only for some combinations of inputs: the desired value for $m(x, y, z)$, where x, y, and z are all distinct, is not determined. As a result, there are only 15 fitness cases in our *Malcev* tasks, the lowest of all considered benchmarks.

a_1	0	1	2
0	2	1	2
1	1	0	0
2	0	0	1

a)

$$t^A(x, y, z) = \begin{cases} x & if \ x \neq y \\ z & if \ x = y \end{cases}$$

b)

$$m(x, x, y) = m(y, x, x) = y$$

c)

Performance. Table 1 reports the success rates of particular algorithms, resulting from 30 runs of each configuration on every benchmark. The methods clearly fair differently on particular benchmarks. To provide an aggregated perspective on performance, we employ the Friedman's test for multiple achievements of multiple subjects [9]. Compared to ANOVA, it does not require the distributions of variables in question to be normal.

Friedman's test operates on average ranks, which for the considered methods are as follows, for $|P| = 500$ (left) and $|P| = 1000$ (right):

DOC	IFS	RAND	GP	DOC	IFS	RAND	GP
1.93	2.20	2.50	**3.36**	1.76	2.33	2.60	**3.30**

The *p*-value for Friedman test is $\ll 0.001$, which strongly indicates that at least one method performs significantly different from the remaining ones. We conducted post-hoc analysis using symmetry test [7]: bold font marks the methods that are outranked at 0.05 significance level by the *first* method in the ranking.

Analysis. Although DOC ranks first for both population sizes, it does not seem to be much better than IFS, a substantially simpler method. We hypothesize that this may be an effect of *overspecialization*, which may be likened to *focusing*, one of so-called coevolutionary pathologies [5, 23]. Even though evolving a program

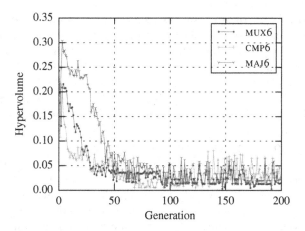

Fig. 2. Average hypervolume of programs in population across generations for the Mux6, Cmp6 and Maj6 benchmarks.

that passes all tests is hard, it may be relatively easy to find programs that perform well on a certain *subset* of tests while failing the other tests. For instance, in the Boolean benchmark *Cmp6*, the task is to determine whether the number encoded by the three least significant input bits b_0, b_1, b_2 is smaller than the number encoded by the three most significant bits b_3, b_4, b_5. A program that checks if b_2 is off and simultaneously b_5 is on solves the quarter of $2^6 = 64$ tests in this task. This can be expressed with a mere few instructions from the assumed instruction set, e.g., as $(b_2 \ nor \ b_2) \ and \ b_5$. It is possible that evolution exploits this opportunity by synthesizing programs that focus on such easy subproblems.

To verify this hypothesis, we define the *hypervolume* of program's performance as characterized by the k derived objectives $o_1, \ldots, o_k$, i.e.,

$$h(p) = \prod_{i=1}^{k} o_i(p). \tag{4}$$

The key property of hypervolume is that it increases as the scores on o_is become more balanced. Consider two programs p_1, p_2 with the same overall fitness, i.e., $\sum_i o_i(p_1) = \sum_i o_i(p_2)$. Assume the scores of p_1 on o_is vary, while those of p_2 are all the same, i.e., $o_i(p_2) = \sum_i o_i(p_1)/k$. In such a case, $h(p_2) > h(p_1)$. $h(p_2)$ is the maximum hypervolume for all possible distributions of the same scalar fitness across the derived objectives.

Figure 2 plots the hypervolume of programs in population across generations for the *Mux*6, *Cmp*6 and *Maj*6 benchmarks, averaged over population and over 90 evolutionary runs. We observe dramatic decline of this measure with evolution time. With the other benchmarks exhibiting similar characteristics, we can conclude that indeed the programs evolved by DOC tend to overspecialize.

Promoting uniform progress. The NSGA-II selection procedure operates on Pareto ranks and as such is agnostic to a more detailed location of a given point in the multiobjective space that spans o_is. As long as two programs have the same Pareto rank, they will be equally valuable (unless differentiated by sparsity). This holds even if one of them is on the very extreme of Pareto front, i.e., attains zero value of one or more objectives. In other words, NSGA-II lacks mechanisms that would promote achieving balanced performance on *all* derived objectives simultaneously.

This observation, combined with the above demonstration of overspecialization, immediately points to a remedy. If hypervolume is a natural measure of balanced performance on all objectives, why not use it as a search driver? To verify this idea, we come up with a straightforward variant of DOC, called **DOC-P** in the following. DOC-P aggregates the scores on derived objectives using Formula 4, and uses the resulting hypervolume as fitness in combination with tournament selection of size 7, as in the other control configurations.

We also propose a second variant of this idea, **DOC-P**, which additionally *weights* the objectives by the number of tests (columns in G) included in each objective, i.e.,

$$h_D(p) = \prod_{i=1}^{k} |T_i| o_i(p). \tag{5}$$

In effect, $h_D(p)$ is based on the *number* of tests passed by p on each derived objectives, while h relied on the raw values of o_j, i.e., *mean* test outcomes in clusters.

The columns in Table 1 marked DOC-P and DOC-D report the results of these methods. Below, we present the average ranks of all methods, including these extensions:

DOC-D	DOC-P	IFS	DOC	RAND	GP		DOC-P	DOC-D	DOC	IFS	RAND	GP
1.70	2.43	3.56	**3.63**	**4.33**	**5.33**		2.20	2.43	3.10	3.66	**4.50**	**5.10**

We observe both setups dramatically improving the performance compared to the original DOC. For $|P| = 500$ (left), the DOC-D ranks the best, outperforming GP, RAND and the multiobjective variant of DOC in a statistically significant way. The difference is statistically insignificant for IFS, but both DOC-D and DOC-P score higher success rates more often and manage to solve two problems that remained unsolved by other algorithms, i.e., *Disc4* and *Disc5*.

The result are quite similar when $|P| = 1000$ (right), however this time DOC-P stands out as the best, albeit its rank is only slightly higher than that of DOC-D. Larger population is also beneficial for multiobjective DOC allowing it to achieve lower rank than IFS and beat GP in a statistically significant way. We speculate that this effect is directly related to the Pareto-fronts becoming densely populated, and thus decreasing the risk of over-specialization.

The experimental results clearly indicate that both DOC-P and DOC-D are more likely to find an ideal solution than the traditional GP and prove capable of solving problems that GP struggles with. If a larger population size is admissible, multiobjective DOC also emerges as a viable alternative to IFS and conventional GP.

6 Conclusions

In this paper we proposed a method that heuristically derives new search objectives by clustering the outcomes of interactions between the programs in population and the tests. The derived search objectives, either combined with the NSGA-II or combined into a hypervolume of program's performance, effectively enhance conventional GP. DOC manages to produce a low number of objectives that approximately capture the capabilities of evolving programs. Once identified, DOC maintains the presence of such skills in the population, even if the programs featuring them are inferior according to the conventional fitness. In this study, the capabilities in question concerned program output; in general, they may correspond to program *behaviors* in a broader sense, or reflect whether they satisfy certain *conditions*. Such generalizations deserve investigation in the future work.

When seen from the perspective of the overall evolutionary workflow, DOC broadens the 'bottleneck of evaluation' described in Introduction in characterizing the candidate solutions with multiple objectives rather than with a single one. Objectives derived by DOC constitute alternative search drivers that replace the conventional fitness function and guide the search in a single- or multiobjective fashion. Ultimately, capabilities elaborated by particular individuals have the chance of being fused in their offspring and so ease reaching the search goal. In this context, there is an interesting relationship between the derived objectives and the *intermediate* results produced by programs studied in behavioral evaluation [12] and pattern-guided genetic programming [13].

Acknowledgments. P. Liskowski acknowledges support from grant no. 09/91/DSPB/ 0572.

References

1. Bucci, A., Pollack, J.B., de Jong, E.: Automated extraction of problem structure. In: Deb, K., Tari, Z. (eds.) GECCO 2004. LNCS, vol. 3102, pp. 501–512. Springer, Heidelberg (2004)
2. de Jong, E.D., Bucci, A.: DECA: dimension extracting coevolutionary algorithm. In: Cattolico, M., et al., (eds.) GECCO 2006: Proceedings of the 8th Annual Conference on Genetic and Evolutionary Computation, pp. 313–320. ACM Press, Seattle, Washington, USA (2006)
3. de Jong, E.D., Pollack, J.B.: Ideal evaluation from coevolution. Evol. Comput. **12**(2), 159–192 (2004)
4. Deb, K., Pratap, A., Agarwal, S., Meyarivan, T.: A fast and elitist multiobjective genetic algorithm: NSGA-II. IEEE Trans. Evol. Comput. **6**(2), 182–197 (2002)

5. Ficici, S.G., Pollack, J.B.: Challenges in coevolutionary learning: arms-race dynamics, open-endedness, and mediocre stable states. In: Proceedings of the Sixth International Conference on Artificial Life, pp. 238–247. MIT Press (1998)

6. Ficici, S.G., Pollack, J.B.: Pareto optimality in coevolutionary learning. In: Kelemen, J., Sosík, P. (eds.) ECAL 2001. LNCS (LNAI), vol. 2159, p. 316. Springer, Heidelberg (2001)

7. Hollander, M., Wolfe, D.A., Chicken, E.: Nonparametric Statistical Methods, vol. 751. John Wiley & Sons, Weinheim (2013)

8. Jaśkowski, W., Krawiec, K.: Formal analysis, hardness and algorithms for extracting internal structure of test-based problems. Evol. Comput. **19**(4), 639–671 (2011)

9. Kanji, G.K.: 100 Statistical Tests. Sage, London (2006)

10. Krawiec, K., Lichocki, P.: Using co-solvability to model and exploit synergetic effects in evolution. In: Schaefer, R., Cotta, C., Kołodziej, J., Rudolph, G. (eds.) PPSN XI. LNCS, vol. 6239, pp. 492–501. Springer, Heidelberg (2010)

11. Krawiec, K., O'Reilly, U.M.: Behavioral programming: a broader and more detailed take on semantic GP. In: Igel, C. (ed.) GECCO 2014: Proceedings of the 2014 Conference on Genetic and Evolutionary Computation, pp. 935–942. ACM, Vancouver, BC, Canada, 12–16 July 2014

12. Krawiec, K., O'Reilly, U.-M.: Behavioral search drivers for genetic programing. In: Nicolau, M., Krawiec, K., Heywood, M.I., Castelli, M., García-Sánchez, P., Merelo, J.J., Rivas Santos, V.M., Sim, K. (eds.) EuroGP 2014. LNCS, vol. 8599, pp. 210–221. Springer, Heidelberg (2014)

13. Krawiec, K., Swan, J.: Pattern-guided genetic programming. In: Blum, C. (ed.) GECCO 2013: Proceeding of the Fifteenth Annual Conference on Genetic and Evolutionary Computation Conference, pp. 949–956. ACM, Amsterdam, The Netherlands, 6–10 July 2013

14. Lasarczyk, C.W.G., Dittrich, P., Banzhaf, W.: Dynamic subset selection based on a fitness case topology. Evol. Comput. **12**(2), 223–242 (2004)

15. Liskowski, P., Krawiec, K.: Discovery of implicit objectives by compression of interaction matrix in test-based problems. In: Bartz-Beielstein, T., Branke, J., Filipič, B., Smith, J. (eds.) PPSN 2014. LNCS, vol. 8672, pp. 611–620. Springer, Heidelberg (2014)

16. McKay, R.I.B.: Committee learning of partial functions in fitness-shared genetic programming. In: Industrial Electronics Society, 2000. IECON 2000. 26th Annual Conference of the IEEE Third Asia-Pacific Conference on Simulated Evolution and Learning 2000. vol. 4, pp. 2861–2866. IEEE Press, Nagoya, Japan, 22–28 October 2000

17. McKay, R.I.B.: Fitness sharing in genetic programming. In: Whitley, D., Goldberg, D., Cantu-Paz, E., Spector, L., Parmee, I., Beyer, H.G. (eds.) Proceedings of the Genetic and Evolutionary Computation Conference (GECCO-2000), pp. 435–442. Morgan Kaufmann, Las Vegas, Nevada, USA, 10–12 July 2000

18. Noble, J., Watson, R.A.: Pareto coevolution: using performance against coevolved opponents in a game as dimensions for pareto selection. In: Spector, L., et al., (eds.) Proceedings of the Genetic and Evolutionary Computation Conference (GECCO-2001), pp. 493–500. Morgan Kaufmann, San Francisco, California, USA, 7–11 July 2001

19. Pelleg, D., Moore, A.W., et al.: X-means: extending k-means with efficient estimation of the number of clusters. In: ICML, pp. 727–734 (2000)

20. Smith, R.E., Forrest, S., Perelson, A.S.: Searching for diverse, cooperative populations with genetic algorithms. Evol. Comput. **1**(2), 127–149 (1993)

21. Spector, L., Clark, D.M., Lindsay, I., Barr, B., Klein, J.: Genetic programming for finite algebras. In: Keijzer, M. (ed.) GECCO 2008: Proceedings of the 10th Annual Conference on Genetic and Evolutionary Computation, pp. 1291–1298. ACM, Atlanta, GA, USA, 12–16 July 2008

22. Tomassini, M., Vanneschi, L., Collard, P., Clergue, M.: A study of fitness distance correlation as a difficulty measure in genetic programming. Evol. Comput. **13**(2), 213–239 (2005)

23. Watson, R.A., Pollack, J.B.: Coevolutionary dynamics in a minimal substrate. In: Spector, L., et al., (eds.) Proceedings of the Genetic and Evolutionary Computation Conference (GECCO-2001), pp. 702–709. Morgan Kaufmann, San Francisco, California, USA, 7–11 July 2001

Evolutionary Design of Transistor Level Digital Circuits Using Discrete Simulation

Vojtech Mrazek[(✉)] and Zdenek Vasicek

Faculty of Information Technology, Brno University of Technology,
Božetěchova 2, 612 66 Brno, Czech Republic
{imrazek,vasicek}@fit.vutbr.cz

Abstract. The objective of the paper is to introduce a new approach to the evolutionary design of digital circuits conducted directly at transistor level. In order to improve the time consuming evaluation of candidate solutions, a discrete event-driven simulator was introduced. The proposed simulator operates on multiple logic levels to achieve reasonable trade-off between performance and precision. A suitable level of abstraction reflecting the behaviour of real MOSFET transistors is utilized to minimize the production of incorrectly working circuits. The proposed approach is evaluated in evolution of basic logic circuits having more than 20 transistors. The goal of an evolutionary algorithm is to design a circuit having the minimal number of transistors and exhibiting the minimal delay. In addition to that, various parameter settings are investigated to increase the success rate of the evolutionary design.

Keywords: Evolutionary design · Transistor-level · Digital circuits · Cartesian genetic programming

1 Introduction

In recent years, a lot of papers showing the merits of evolutionary design techniques in the field of digital circuits design have been published. Implementation of various combinational circuits competitive to the circuits designed using conventional approaches have been obtained by using cartesian genetic programming (CGP) which is considered to be the most efficient technique to perform the gate-level evolutionary design [2–4,7].

However, while the gate-level evolutionary design represents an intensively studied research area, the synthesis of transistor-level digital circuits remains, in contrast with design of transistor-level analog circuits, on a peripheral concern of the researchers despite the fact that even some basic logic expressions can be implemented much effectively at transistor level. Only few papers were devoted to evolution of digital circuits directly at transistor level. Zaloudek et al. published an approach based on a simple simulator which was designed to quickly evaluate the candidate solutions [11]. Unfortunately, a rough approximation of transistor behavior caused that this approach produced many incorrectly working circuits. Trefzer used another technique to evolve some basic logic gates [6].

© Springer International Publishing Switzerland 2015
P. Machado et al. (Eds.): EuroGP 2015, LNCS 9025, pp. 66–77, 2015.
DOI: 10.1007/978-3-319-16501-1_6

Instead of using a time consuming analog circuit simulator, a reconfigurable analog transistor array was employed. However, it was shown that many of the discovered solutions relied on some properties of the utilized reconfigurable array. About 50 % of the evolved circuits failed in the analog simulation. Walker et al. used a different technique to evolve transistor-level circuits [8]. In order to speed up the time consuming evaluation of candidate solutions, a cluster of SPICE-based simulators was utilized. Even if it was possible to evolve correct solutions, only small problem instances could be investigated due to the overhead of SPICE simulators.

A new approach to the evolutionary design of digital circuits is introduced in this paper. In this work, the evolutionary approach operates directly at transistor level. Since the evolutionary-based approach requires generating a large number of candidate solutions, it is necessary to minimize the time needed to evaluate the candidate circuits in order to obtain a satisfactory success rate. However, a reasonable level of abstraction must be applied to avoid production of incorrectly working circuits. In order to address this issue, a discrete simulator which operates on multiple logic levels is proposed. It is expected that this approach enables to achieve reasonable trade-off between performance and precision.

The paper is organized as follows. Section 2 discusses behavior of real unipolar transistors. Section 3 introduces the proposed method. Section 4 summarizes and analyses the obtained results. The analysis of the discovered circuits is performed using a SPICE simulator. Finally, concluding remarks are given in Sect. 5.

2   Behavior of MOSFET Transistors

Behavior of the MOSFET transistors can be described at various levels of abstraction.

At the most accurate level, transistor circuits are modeled using a complex system of equations having tens of parameters that are derived from the underlying device physics. In order to accurately simulate the transistor level circuits, SPICE-based simulators are usually used. Apart from the commercial simulators such as HSPICE or PSPICE, there exist also academic tools such as ngSPICE. Even if the SPICE-based simulators provide a wide variety of MOS transistor models with various trade-offs between complexity and accuracy, the runtime grows rapidly with the increasing size of the simulated circuits. To reduce the time of simulation, a multithreaded version of SPICE simulator or an FPGA-based hardware acceleration can be utilized [1].

On the other hand, so-called switch-level model can be used [10]. A switch-level simulator models MOS circuits using a network of transistors acting as switches. Each transistor can be in one of three discrete states – open, closed or unknown. Compared to the SPICE-based simulators, the speed of the simulation is improved in orders of a magnitude. This model can acquire aspects that cannot be expressed at gate model, however, the accuracy is naturally lower compared to the approach mentioned in the previous paragraph. For example, the value of threshold voltage influencing state of the transistors is completely ignored. Moreover, the accuracy of simulation decreases as the transistors shrinks.

2.1   Discrete Model Suitable for Evolutionary Design

As it was discussed in the Sect. 1, a fast simulator is needed to enable the evolutionary design to sufficiently explore the search space. Simultaneously, reasonable accuracy is required to evolve the correct circuits that will work in real
environment. In order to meet these requirements, we propose to utilize a discrete simulator which exhibits speed of the switch-level simulators and accuracy
of the SPICE-based simulators. We propose to use a model (abstracted from
dynamic parameters such as power consumption or delay) based on the switch-
level transistor model extended to a threshold drop degradation effect.

Threshold voltage, commonly abbreviated as V_{tp} (V_{tn}), is the minimum gate-
to-source voltage differential that is needed to create a conducting path between
the source and drain terminals. As a consequence of the threshold voltage, degraded
voltage values can be presented in MOS circuits. An open n-MOS transistor is
known to pass logic 0 (i.e. V_{ss}) well but logic 1 (i.e. V_{dd}) poorly. This loss is known as
threshold drop. An attempt to pass logic 1 never gives value above $V_{dd} - V_{tn}$. Similarly, p-MOS transistor is known to pass logic 0 poorly. The reduction in voltage
swing can be beneficial to the power consumption. The designer has to be careful,
however, because the degraded output may cause circuit malfunction.

As a target technology, TSMC with feature size equal to $0.25\,\mu m$ is chosen.
The following parameters of p-MOS and n-MOS transistors will be utilized in
MOS circuits. The length of the n-MOS transistor is $L_N = 0.25\,\mu m$, width is
$W_N = 0.5\,\mu m$. p-MOS transistors have $L_P = 0.25\,\mu m$ and $W_P = 2W_N = 1\,\mu m$.

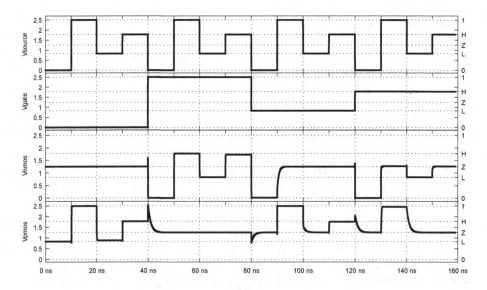

Fig. 1. Output waveforms for p-MOS and n-MOS transistors for various voltage applied
to the source and gate terminals. The waveform was obtained using an analog SPICE
simulator, a TSMC $0.25\,\mu m$ technology and 2.5 V power supply. The corresponding
discrete values are shown on the right side.

According to the simulation, $V_{tn} = 0.987\ V$ and $V_{tp} = 0.717\ V$. In order to support various implementations of digital circuits, we will distinguish among six voltage levels: logic 0 (denoted as '0'), logic 1 ('1'), degraded 0 (V_{tp}, 'L'), degraded 1 ($V_{dd} - V_{tn}$, 'H'), high impedance ('Z') and undefined value ('X'). A SPICE-based simulator was used to derive the discrete model. The results of simulation are given in Fig. 1. The fourth terminal of p-MOS (n-MOS) is connected to V_{dd} (V_{ss}). In order to detect high impedance state, outputs of p-MOS and n-MOS transistors are connected to a voltage divider.

Let us discuss behavior of n-MOS transistor (p-MOS works analogically). If logic 0 is applied to the gate, the transistor is closed and its output is in a high impedance state. The similar situation occurs if 'L' is used. However, if $V_{gate} = $ 'L' and $V_{source} = $ '0', the transistor is not completely closed. As we do not want to model strength of the signal values, we need to suppose that the output is in a high impedance state. This little inaccuracy does not constitute any serious problem due to the presence of stronger values within a circuit. If logic 1 or 'H' is applied to the gate, the transistor is open. Logic 0 as well as 'L' connected to the source are fully transferred to output, but logic 1 and 'H' are degraded. As we can see, the double degraded value can not be recognized from high impedance state. Hence, we have to avoid the double degradations that may cause malfunctions.

The behaviour of n-MOS and p-MOS transistors which follows the results obtained from the SPICE-based simulation valid for the chosen technology and power supply is summarized in Table 1.

Table 1. Behavior of n-MOS and p-MOS transistors modeled using six discrete values.

	n-MOS							p-MOS					
gate	source						gate	source					
	1	H	L	0	Z	X		1	H	L	0	Z	X
1	H	X	L	0	Z	X	1	Z	Z	Z	Z	Z	X
H	X	X	L	0	Z	X	H	Z	Z	Z	Z	Z	X
L	Z	Z	Z	Z	Z	X	L	1	H	X	X	Z	X
0	Z	Z	Z	Z	Z	X	0	1	H	X	L	Z	X
Z	Z	Z	Z	Z	Z	X	Z	Z	Z	Z	Z	Z	X
X	X	X	X	X	X	X	X	X	X	X	X	X	X

3   The Proposed Method

3.1   Circuit Representation

In order to evolve complex digital circuits at the transistor level a suitable representation enabling to encode bidirectional graph structures containing junctions is needed. To address this problem, we proposed an encoding inspired by CGP [2].

Each digital circuit having n_i primary inputs and n_o primary outputs (i.e. a candidate solution) is represented using an array of nodes arranged in n_c columns and n_r rows. Each node consists of two source terminals and one output terminal. Each node can act as p-MOS transistor, n-MOS transistor, or junction. The utilized nodes are shown in Fig. 2. Source terminals of each node can independently be connected to the output terminal of a node placed in previous l columns. In addition to that, source terminals of any transistor node can be connected to one of the primary circuit inputs.

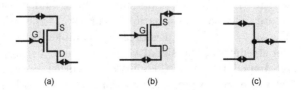

(a)   (b)   (c)

Fig. 2. Basic building blocks of transistor-level circuits: (a) p-MOS transistor, (b) n-MOS transistor, and (c) junction that combines two signals together. If a proper voltage is applied on the gate electrode denoted as G (V_{ss} for p-mos, V_{dd} for n-mos), transistor connects its source electrode (denoted as S) with drain (D). Possible directions of signal flow which have to be considered during the evaluation are shown.

Presence of the junction node represents the main feature of the proposed technique. This node is able to combine two input signals and one output signal together. As a consequence of that, loops and multiple connections are natively supported.

The following encoding scheme is utilized. The primary inputs and node outputs are labeled from 0 to $n_i + n_c \cdot n_r - 1$. A candidate solution is represented in the chromosome by $n_c \cdot n_r$ triplets (x_1, x_2, f) determining for each node its function f, and label of nodes x_1 and x_2 connected to the source terminals. Apart from that, negative indices $-2 - n_i < x_i \leq -2$ are allowed in case of x_i. The negative value indicates that the inverted primary variable labeled as $|x_i|$ is required. The last part of the chromosome contains n_o integers specifying the labels of nodes where the n_o primary outputs are connected to. The first two primary inputs are reserved for power supply rails.

Figure 3 demonstrates the principle of utilized encoding on a XNOR circuit implemented using pass-transistor logic. The shown chromosome encodes a candidate circuit using eight nodes, however, only some of them contribute to the phenotype and are active.

3.2   Evaluation of the Candidate Solutions

Evaluation of the candidate solutions encoded using the proposed representation consists of two steps.

Firstly, set of active nodes is determined. Only the active nodes are considered during the evaluation. The inactive nodes are ignored. Potentially unwanted nodes

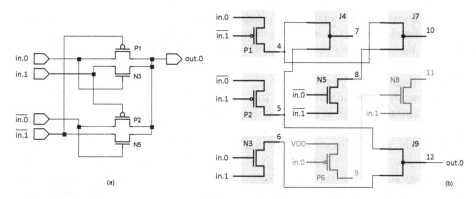

Fig. 3. Example of a candidate circuit implementing function XNOR using eight transistors (four transistors are used to implement inverted variables $\overline{in.0}$ and $\overline{in.1}$). Parameters are as follows: $n_i=4$ $(0,V_{dd},in.0,in.1)$, $n_o=1$ $(out.0)$, $n_c=3$, $n_r=3$, $l=2$. Chromosome: $(2,-3,pmos)(-2,3,pmos)(3,2,nmos)(4,5,junction)$ $(-3,-2,nmos)(1,2,pmos)(4,8,junction)(9,3,nmos)(5,6,junction)(12)$.

causing short-circuits can be removed in this step. A node is active if either (a) its output is connected to any of the primary outputs, or (b) it is a transistor node and its output is connected to the source of an active node, or (c) it is a junction node whose source terminal is connected to an active node. The detection of active nodes can be performed in linear time complexity.

Then, multi-level discrete event-driven simulator is utilized to determine response for each input combination. The advantage of this approach is that only necessary nodes are updated if there is a change of a value. The following steps are used to determine output value of for a given input combination. Firstly, outputs of all nodes are initialized to the value 'Z'. Then, value 0 and 1 are assigned to the first two primary inputs. This change triggers re-evaluation of all the nodes connected directly to the power supply rails. Each node determines its new output value and propagates it to all related nodes. As an open transistor connect source with drain, bidirectional data-flow have to be utilized. It means that the new value must be propagated to the nodes connected not only to the drain but also to source terminal. Similarly, junctions have to propagate the new value to all terminals. The new value of a junction node is calculated as the strongest value presented on all the terminals. The new value of a transistor node is determined according to the value connected to the source as well as drain. During the evaluation of a new output value of a transistor node, the new calculated value is compared with current value at drain terminal. If the values are not compatible, short circuit exception is raised. Otherwise, the stronger value is propagated to all related nodes. The relation between the discrete values is as follows: 'Z' ≺ 'L' ≺ '0' ≺ 'X'; 'Z' ≺ 'H' ≺ '1' ≺ 'X'. It means that if at least one of the values is equal to 'X', 'X' is propagated to all related nodes.

Each transistor has associated a state which determines whether the transistor is in direct or reverse mode. The current flows from drain to source in reverse mode. It happens when 'Z' is assigned to source terminal and a value different from 'Z' is connected to the drain. This state helps to avoid situation in which a double degradation could happen.

In order to avoid malfunction circuits, final test is performed at the end of the simulation. If there is at least a single transistor with 'Z' state assigned to its gate terminal, short circuit exception is raised.

The principle of discrete simulation will be demonstrated for $in_0 = 1$ and $in_1 = 0$ using the candidate circuit shown in Fig. 3. The primary inputs are successively initialized to the following values: V_{dd} (i.e., the primary input with index 0) ← '1', $V_{ss}(1)$ ← '0', $in_0(2)$ ← '1', $in_1(3)$ ← '0'. Then the inverted values are assigned $\overline{in_0}(-2)$ ←'0', $\overline{in_1}(-3)$←'1' As no power rail is used in the example, the first two assignments do not trigger any reevaluation. However, assignment of value '1' to in_0 causes that P1 and N3 are evaluated. Nor P1 nor N3 have fully specified inputs, thus these changes do not generate any new event. In the next step, in_1 connected to P2 and N3 is assigned. Now, the node N3 has fully specified inputs and the new calculated value '0' is propagated through drain to the node J9. Then, the value of $\overline{in_0}$ is changed to '0'. As a consequence of that, P2 is evaluated to 'L' and propagated through J4 to J9. In addition to that, N5 is refreshed. Because there is a stronger value, '0', assigned to the other pin of J9, the '0' is propagated back to the output terminal of transistor P2 and junction J4. The, $\overline{in_1}$ becomes to be logical '1'. Transistor P1 is closed, so the drain is in high impedance state. This value is propagated to J4, however '0' presented at the second terminal is stronger and it is propagated back to P1 and then to J7. The last transistor which has to be evaluated is the closed transistor N5 with 'Z' at its output. High impedance state is delivered to J7, but J7 already contains a stronger value '0'. Primary output is connected to the node J9 which has value '0' on its output. This value corresponds with the XNOR specification, so the circuit produces a valid output for the used input vector.

3.3 Search Strategy

As a search algorithm, $(1 + \lambda)$ evolutionary strategy is utilized [2]. The initial population is randomly generated. Every new population consists of the best individual and λ offspring created using a point mutation operator which modifies h randomly selected genes. In the case when two or more individuals have received the same fitness score in the previous population, the individual which did not serve as a parent in the previous population will be selected as a new parent. This strategy is used to ensure the diversity of population. The evolution is terminated when a predefined number of generations is exhausted or a required solution is found.

The search is guided by the fitness function which determines how good the current candidate circuit is. For evolution of logic circuits, all possible input combinations have to be applied at the candidate circuit inputs. The output values are collected and the goal is to minimize the difference between obtained responses and required Truth table. In order to smooth the search space, the fitness value is constructed as follows. If an obtained output value equals to the expected one, 5 points are added to the fitness value. If the calculated value exhibits the same polarity but represents degraded voltage, 2 points are used. Otherwise, no point is added because the response is invalid. Additional penalties may be applied. If there is a short-circuit exception asserted during the simulation, the simulation is terminated and penalty is applied to the total fitness value. Similarly, if the simulator exceeds the predefined number of steps (i.e. node outputs are not in stable state), the simulation is terminated and the fitness value is penalized. As soon as a fully working solution is found, the number of utilized transistors is reduced. Two points are added for each unused node and one point for node which acts as junction. Note that the transistors required to implement inverted input of the utilized variables are considered.

4   Experimental Results

The proposed method was evaluated in the evolution of basic logic circuits as well as some benchmark circuits whose conventional solutions consist of up to 30 transistors. In particular, we tried to evolve XOR and XNOR gate, 3 bit majority, 1 bit full adder and benchmark circuits b1, c17, newtag, mc, daio and lion from *LGSynth benchmarks*. The goal of the experiments was to evolve fully functional implementations exhibiting full voltage swing on the outputs.

In order to investigate the effect of array size, three arrangements are used for each benchmark circuit. The first two configurations utilize a single row of nodes, while the third uses an array consisting of two rows. The total number of nodes was chosen according to the number of transistors required to implement a given function using a conventional design approach.

In addition to that, the impact of various connection possibilities was investigated. Firstly, the presence of inverted input variables introduced in Sect. 3.1 and its impact on the success-rate was studied. Then, additional restriction to the connection of source terminal of p-MOS and drain terminal of n-MOS was applied. We prevent to connect this electrode directly to the primary inputs. As a consequence of that, implementations with higher operating frequency can be evolved. This setup is denoted as 'S/D←N', while the unrestricted setup is denoted as 'S/D←I+N'.

The results were obtained from 20 independent runs using the following experimental setup: $\lambda = 4, l = n_c, h = 5$. The evolution is terminated after 8 h or when no improvement was achieved within the last hour. All the successfully evolved solutions were validated using a SPICE simulator.

The results were compared with a reference implementation described at gate-level and implemented using standard cells.

The impact of the introduced restriction and the presence of implicit inverters is evaluated by means of a *success proportion* [9]. Success proportion is the cumulative probability of success calculated by the number of runs that have found a solution at or before generation i divided by the total number of runs

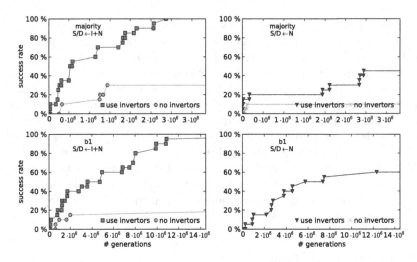

Fig. 4. Success proportion of the evolutionary design of 'majority' and 'b1' benchmark circuits. The array consisting of a single row and 30 columns for 'majority' and 60 columns for 'b1' are used.

in the experiment. A successful run is such a run in which a fully working solution was discovered. The results for two chosen benchmark circuits are given in Fig. 4. As it can be seen, the usage of implicit inverters significantly increased the performance of the evolutionary design. On the other hand, the restriction applied to the source (drain) terminals of p-MOS (n-MOS) nodes reduce the performance of the evolution. Substantially higher number of generations are needed to achieve the same success rate.

The success rate of the evolutionary design for the chosen digital circuits is summarized in Table 2. In addition to that, we analyzed the evolved solutions and determined the number of utilized transistors (see the last two columns). Similarly to the previous findings, the usage of implicit inverters as well as the unrestricted possibilities of S/D terminal connections improved the performance of the evolutionary approach in all cases. Another parameter which can have a great impact on the success rate is the size of array. Too small array on the one hand and too large array on the other hand have a negative impact on the success rate. While the small array may prevent to find a valid solution because there is not a space to represent a target circuit, large array increases substantially the search space. Fortunately, it seems that increasing of the number of available nodes does not increase the size of the evolved circuit.

The discovered circuits were verified and characterized using a SPICE simulator with an accurate transistor model. Except of a single evolved implementation of 'b1' circuit, all the circuits were valid and operated correctly. Thus we can conclude that the proposed discrete abstraction is successful.

Table 3 summarizes the basic parameters of the evolved solutions and the conventional implementations. Apart from the number of utilized transistors,

Table 2. Success rate for the benchmark circuits for various array sizes, connection possibilities and availability of inverted primary inputs.

	$n_r \times n_c$	S/D←N+I		S/D←N		# transistors	
		with inv.	w/o inv.	with inv.	w/o inv	min	max
xnor	1 × 10	100 %	65 %	0 %	0 %	6	8
	1 × 15	100 %	100 %	100 %	5 %	6	12
	2 × 15	100 %	100 %	100 %	45 %	6	12
xor	1 × 10	100 %	75 %	0 %	0 %	6	8
	1 × 15	100 %	100 %	100 %	5 %	6	12
	2 × 20	100 %	100 %	100 %	5 %	6	12
majority	1 × 20	100 %	25 %	0 %	5 %	10	14
	1 × 30	100 %	30 %	45 %	10 %	10	16
	2 × 30	80 %	35 %	60 %	15 %	10	17
adder-1	1 × 30	30 %	5 %	0 %	0 %	14	20
	1 × 40	65 %	0 %	0 %	0 %	18	20
	2 × 40	50 %	0 %	5 %	0 %	18	25
b1	1 × 40	100 %	15 %	40 %	0 %	12	19
	1 × 60	100 %	20 %	60 %	0 %	12	20
	2 × 60	75 %	5 %	25 %	0 %	12	23
c17	1 × 40	5 %	0 %	0 %	0 %	22	24
	1 × 60	5 %	0 %	0 %	0 %	25	26
	2 × 60	0 %	0 %	5 %	0 %	25	28

delay and maximum operating frequency is given. If we compare the maximum operating frequency of the evolved circuits with the conventional circuits, we can see a significant improvement in all cases except the circuit 'c17'. This result is very encouraging, because the delay was not optimized explicitly. We analyzed the circuits and determined that this improvement was achieved by replacing traditional gates implemented as CMOS logic with much effective implementation which utilized so-called transmission-gates. The usage of transmission-gates increases the speed but simultaneously reduces the number of utilized transistor.

A lot of different implementations were discovered. Example of an evolved circuit of one bit adder is shown in Fig. 5. The discovered circuit is similar to low-power full adder consisting of 14 transistors which was introduced in [5]. The evolution was able to discover an implementation which belongs to the family of pass-transistor logic. The evolved solution utilizes three transmission gates to provide fast and compact solution and exhibits approx. 27 % reduction in power consumption compared to the common CMOS implementation. Carry is represented by output labeled as out_0 and sum is available at out_1. Input in_2 corresponds to the input carry.

Table 3. Parameters of the conventional as well as evolved digital circuits. The first part of the table contains the number of inputs, number of outputs and time and number of generations required to evolve the solution. Then, the parameters of conventional implementation are given. (a) Contains parameters of the fastest discovered solution, while (b) contains parameters of the most compact evolved solution.

		xor	xnor	majority	adder-1	b1	c17
	Inputs	2	2	3	3	3	5
	Outputs	1	1	1	2	2	2
	Time of evolution (min)	10	10	10	120	60	480
	Max. # generations	$14 \cdot 10^6$	$14 \cdot 10^6$	$5 \cdot 10^6$	$45 \cdot 10^6$	$30 \cdot 10^6$	$80 \cdot 10^6$
	Delay (ps)	208.3	180.9	335.2	422.7	360.1	324.0
	Frequency (GHz)	4.80	5.53	2.98	2.37	2.78	3.09
	Transistors	8	8	22	48	30	28
(a)	Delay (ps)	87.5	87.8	271.4	291.4	173.2	355.4
	Frequency (GHz)	11.43	11.39	3.68	3.43	5.77	2.81
	Transistors	6	8	16	14	16	24
(b)	Delay (ps)	87.5	142.4	599.3	291.4	401.5	573.8
	Frequency (GHz)	11.43	7.02	1.67	3.43	2.49	1.74
	Transistors	6	6	10	14	12	22

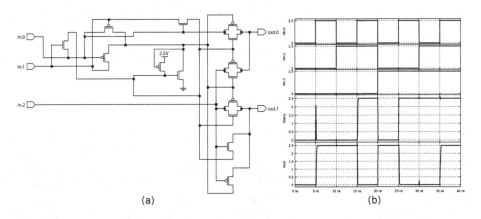

(a)   (b)

Fig. 5. (a) The most compact and simultaneously the fastest circuit consisting of 14 transistors implementing one-bit full adder. (b) Output waveform obtained using a SPICE simulator.

5 Conclusion

A new approach suitable to the evolutionary design of digital circuits conducted directly at transistor level was introduced in this paper. A discrete event-driven simulator operating on multiple logic levels was utilized to achieve reasonable trade-off between performance and precision. The proposed method was evaluated on a set of benchmark circuits. In order to improve the success rate, implicit inverters were introduced to the encoding.

It was demonstrated that the proposed method is able to produce valid solutions despite the fact that a relative simple discrete model of MOS transistors (compared to the complex models used in SPICE-based simulators) was utilized. According to the analysis of the obtained results, we can confirm, that the evolution was able to discover solutions that are based not only on complementary logic but also on pass-transistor logic.

However, future work has to be conducted to improve the scalability of the proposed method. One of the possible directions is to introduce more complex building blocks such as transmission gate.

Acknowlegement. This work was supported by the Czech science foundation project 14-04197S.

References

1. Kapre, N., DeHon, A.: Accelerating spice model-evaluation using fpgas. In: 17th IEEE Symposium on Field Programmable Custom Computing Machines, FCCM 2009, pp. 37–44, April 2009
2. Miller, J.F. (ed.): Cartesian Genetic Programming. Natural Computing Series, 22nd edn. Springer, Berlin (2011)
3. Miller, J.F., Job, D., Vassilev, V.K.: Principles in the evolutionary design of digital circuits - Part I. Genet. Program. Evolvable Mach. **1**(1), 8–35 (2000)
4. Miller, J.F., Job, D., Vassilev, V.K.: Principles in the evolutionary design of digital circuits - Part II. Genet. Program. Evolvable Mach. **1**(3), 259–288 (2000)
5. Shams, A., Bayoumi, M.: A novel high-performance cmos 1-bit full-adder cell. IEEE Trans. Circ. Syst. II: Analog Digit. Signal Process. **47**(5), 478–481 (2000)
6. Trefzer, M.: Evolution of transistor circuits. Ph.D. thesis, Ruprecht-Karls-Universitt Heidelberg (2006)
7. Vassilev, V., Job, D., Miller, J.: Towards the automatic design of more efficient digital circuits. In: Proceedings of the 2nd NASA/DoD Workshop on Evolvable Hardware, pp. 151–160. IEEE Computer Society, Los Alamitos (2000)
8. Walker, J.A., Hilder, J.A., Tyrrell, A.M.: Evolving variability-tolerant CMOS designs. In: Hornby, G.S., Sekanina, L., Haddow, P.C. (eds.) ICES 2008. LNCS, vol. 5216, pp. 308–319. Springer, Heidelberg (2008)
9. Walker, M., Edwards, H., Messom, C.H.: Success effort and other statistics for performance comparisons in genetic programming. In: IEEE Congress on Evolutionary Computation, pp. 4631–4638 (2007)
10. Weste, N.H., Harris, D.: CMOS VLSI Design: A Circuits and Systems Perspective, 3rd edn. Addison-Wesley, Boston (2005)
11. Žaloudek, L., Sekanina, L.: Transistor-level evolution of digital circuits using a special circuit simulator. In: Hornby, G.S., Sekanina, L., Haddow, P.C. (eds.) ICES 2008. LNCS, vol. 5216, pp. 320–331. Springer, Heidelberg (2008)

M3GP – Multiclass Classification with GP

Luis Muñoz[1], Sara Silva[2,3,4(✉)], and Leonardo Trujillo[1]

[1] Tree-Lab, Posgrado En Ciencias de la Ingeniería, Instituto Tecnológico de Tijuana,
Blvd. Industrial Y Av. ITR Tijuana S/N, Mesa Otay C.P.,
22500 Tijuana, BC, Mexico
{lmunoz,leonardo.trujillo}@tectijuana.edu.mx

[2] BioISI – Biosystems and Integrative Sciences Institute, Faculty of Sciences,
University of Lisbon, Lisbon, Portugal
sara@fc.ul.pt

[3] NOVA IMS, Universidade Nova de Lisboa, 1070-312 Lisboa, Portugal

[4] CISUC, Department of Informatics Engineering,
University of Coimbra, Coimbra, Portugal

Abstract. Data classification is one of the most ubiquitous machine learning tasks in science and engineering. However, Genetic Programming is still not a popular classification methodology, partially due to its poor performance in multiclass problems. The recently proposed M2GP - Multidimensional Multiclass Genetic Programming algorithm achieved promising results in this area, by evolving mappings of the p-dimensional data into a d-dimensional space, and applying a minimum Mahalanobis distance classifier. Despite good performance, M2GP employs a greedy strategy to set the number of dimensions d for the transformed data, and fixes it at the start of the search, an approach that is prone to locally optimal solutions. This work presents the M3GP algorithm, that stands for M2GP with multidimensional populations. M3GP extends M2GP by allowing the search process to progressively search for the optimal number of new dimensions d that maximize the classification accuracy. Experimental results show that M3GP can automatically determine a good value for d depending on the problem, and achieves excellent performance when compared to state-of-the-art-methods like Random Forests, Random Subspaces and Multilayer Perceptron on several benchmark and real-world problems.

Keywords: Genetic programming · Classification · Multiple classes · Multidimensional clustering

1 Introduction

Genetic programming (GP) [10] has been used to solve many difficult problems from various domains, an extensive list of noteworthy examples are reviewed in [7]. However, probably the most straightforward formulation for a GP search is to apply it in supervised machine learning problems, particularly symbolic regression and data classification. In particular, this paper is concerned with the

© Springer International Publishing Switzerland 2015
P. Machado et al. (Eds.): EuroGP 2015, LNCS 9025, pp. 78–91, 2015.
DOI: 10.1007/978-3-319-16501-1_7

latter, an area in which a variety of proposals have been developed [3]. Even though GP has been used to achieve state-of-the-art performance in several benchmark problems and real-world scenarios, it has been particularly difficult to use in multiclass problems [5].

In general, for a supervised classification problem some pattern $\mathbf{x} \in \mathbb{R}^p$ has to be classified in one of M classes $\omega_1, \ldots, \omega_M$ using a training set $\mathcal{X}$ of N p-dimensional patterns with a known class label. Then, the goal is to build a mapping $g(\mathbf{x}) : \mathbb{R}^p \to M$, that assigns each pattern $\mathbf{x}$ to a corresponding class ω_i, where g is derived based on the evidence provided by $\mathcal{X}$. In these problems fitness is usually assigned in two general ways. One approach is to use a *wrapper* method, where GP is used as a feature extraction method that performs the transformation $k(\mathbf{x}) : \mathbb{R}^p \to \mathbb{R}^d$, and then another classifier is used to measure the quality of the transformation based on accuracy or another performance measure. The second approach is to use GP to evolve g directly, performing the feature transformation step implicitly. However, current techniques have left room for improvement, such as automatically determining the proper value for d or dealing with multiclass problems (with $M > 2$).

This paper presents an extension of the recently proposed Multidimensional Multiclass Genetic Programming (M_2GP, from now on M2GP) algorithm [5], a wrapper-based GP classifier that effectively deals with multiclass problems by performing a multidimensional transformation of the input data. The M2GP algorithm uses a fixed number of new feature dimensions d, that must be chosen and fixed before the run starts. On the other hand, the algorithm proposed in this paper is able to heuristically determine an appropriate value for d during the run. To achieve this, the algorithm includes specialized search operators that can increase or decrease the number of feature dimensions produced by each tree, and that allow the search to maintain a population of different transformation functions k that construct a different number of new features dimensions. The proposed algorithm is named M3GP, which stands for M2GP with multidimensional populations.

The remainder of this paper is organized as follows. Section 2 briefly reviews previous works related to the present contribution. Section 3 describes the original M2GP algorithm, explaining how it works and referring to its strengths and its major weakness. Section 4 explains the new improved version of the algorithm, M3GP. Section 5 describes the experiments performed, while Sect. 6 reports and discusses all the results obtained. Finally, Sect. 7 concludes and describes some future work.

2   Related Work

Espejo et al. [3] present a comprehensive discussion on GP-based classification methods, while Ingalalli et al. [5] surveys work specifically focused on multiclass classification with GP. Here we briefly address previous works that present a similar goal as the one outlined for M3GP (besides M2GP), highlighting the main differences to the present contribution.

Lit et al. [8] proposed a layered multipopulation approach, where each layer has d populations, and each population produces a single transformation

$k(\mathbf{x}) : \mathbb{R}^p \to \mathbb{R}$, and classification is performed based on a threshold. While each population is evaluated independently, all of them are combined to generate new feature vectors of dimension d, which are given as input to a new layer, and only the final layer has a single population with $d = 1$. For multiclass problems a Euclidean distance classifier is used and results show the method improves the search efficiency and reduces training time. However, the approach does not improve upon the performance of a standard GP classifier, it is not tested on problems with many classes (highest is $M = 3$), and it requires an a priori setting for the number of layers and populations used in each layer.

Another, more closely related work, is the one presented by Zhang and Rockett [12], who propose a multidimensional feature extraction method that uses a similar solution representation to the one used in M2GP and M3GP. However, the authors set a fixed limit on the maximum number of feature dimensions, set to $d = 50$, and initialize the population with trees that use different number of features within this range. Other important difference is that the authors use a multiobjective search process considering class separation and solution size, and do not explicitly consider multiclass problems, instead relying on a hierarchical nesting of binary classifiers.

3   The M2GP Multiclass Classification Method

M2GP is a recent and innovative method of performing multiclass classification with GP [5]. It has shown to be competitive with the best state-of-the-art classifiers in a wide range of benchmark and real-world problems, something that GP had not achieved.

The algorithm and its strengths. The basic idea of M2GP is to find a transformation, such that the transformed data of each class can be grouped into unique clusters. In M2GP the number of dimensions in which the clustering is performed is completely independent from the number of classes, such that high dimensional datasets may be easily classified by a low dimensional clustering, while low dimensional datasets may be better classified by a high dimensional clustering.

In order to achieve this, M2GP uses a representation for the solutions that allows them to perform the mapping $k(\mathbf{x}) : \mathbb{R}^p \to \mathbb{R}^d$. The representation is basically the same used for regular tree-based GP, except that the root node of the tree exists only to define the number of dimensions d of the new space. Each branch stemming directly from the root performs the mapping in one of the d dimensions. The genetic operators are the regular subtree crossover and mutation, except that the root is never chosen as the crossing or mutation node. However, the truly specialized element of M2GP is the fitness function. Each individual is evaluated in the following way:

– All the p-dimensional samples of the training set are mapped into the new d-dimensional space (each branch of the tree is one of the d dimensions).
– On this new space, for each of the M classes in the data, the covariance matrix and the cluster centroid is calculated from the samples belonging to that class.

- The Mahalanobis distance between each sample and each of the M centroids is calculated. Each sample is assigned the class whose centroid is closer. Fitness is the accuracy of this classification (the percentage of samples correctly classified).

Figure 1 shows an example of clustering of a dataset. The original data, regardless of how many features, or attributes, it contains, is mapped into a new 3-dimensional space by a tree whose root note has three branches, each performing the mapping on each of the three axes X, Y, Z. The fact that the data contains three classes is purely coincidental - it could contain any number of classes, regardless of the dimension of the space. On the left, the clustering was obtained by an individual with low accuracy; on the right, the same data clustered by an individual with accuracy close to 100 %. The class centroids are marked with large circles.

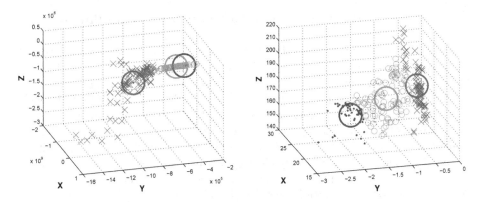

Fig. 1. Example of clustering of a dataset. On the left, clustering obtained by an individual with low accuracy; on the right, the same data clustered by an individual with very good accuracy. The large circles represent the centroids.

At the end of the run, the solution given to the user is composed not only of the tree of the best individual, but also of the respective covariance matrices and cluster centroids. In order to classify unseen data, M2GP uses the tree to map the new samples into the new space, and then uses the covariance matrices and the cluster centroids in order to determine the minimum Mahalanobis distance between each sample and each centroid. (Note that the covariance matrices and cluster centroids are not recalculated when classifying new data). The choice of the Mahalanobis distance instead of the Euclidean distance is not an unnecessary complication of the algorithm, as it allowed a substantial improvement on the quality of the results achieved [5].

M2GP produces trees that are not very large (mean solution size for different problems was reported to range from 24 to 152 nodes [5]), and a higher number of dimensions does not necessarily translate into larger trees.

The weakness. Despite its competitiveness, M2GP suffers from a drawback: how to choose the right number of dimensions for a given problem? M2GP is incapable of adding or removing dimensions during the evolution, so the number of dimensions d is fixed in the beginning of the run. M2GP chooses d based on the observation that the best fitness found in the initial generation is highly correlated with the best fitness found on the final generation [5].

Therefore, before initiating a run, M2GP runs a procedure that iteratively initializes different populations with increasing dimensions (we mean the dimension d mentioned earlier, not the number of individuals in the population) and checks which of these initial populations achieves the best fitness. Starting with $d = 1$, this procedure adds one more dimension and initializes one more population as long as the fitness continues to improve from the previous population. As soon as adding one more dimension degrades fitness, the procedure stops and the dimension yielding the best initial fitness is chosen.

4   M3GP – M2GP with Multidimensional Populations

As described in the previous section, the original M2GP uses a greedy approach to determine how many dimensions the evolved solutions should have. It may happen that by fixing the number of dimensions in the beginning of the run, the algorithm is being kept from finding better solutions during the search, ones that may use a different number of dimensions. In our new improved version, the algorithm evolves a population that may contain individuals of several different dimensions. The genetic operators may add or remove dimensions, and it is assumed that selection will be sufficient to discard the worst ones and maintain the best ones in the population. The next subsections describe M3GP, which stands for M2GP with multidimensional populations.

Initial population. M3GP starts the evolution with a random population where all the individuals have only one dimension. This ensures that the evolutionary search begins looking for simple, one dimensional solutions, before moving towards higher dimensional solutions, which might also be more complex.

For M2GP, a Ramped Half-and-Half initialization [6] skewed to 25 % Grow and 75 % Full was recommended [5], suggesting that a higher proportion of full trees facilitates the initial evolution. Because all the initial M3GP individuals are unidimensional, it makes sense to believe that the need for bigger initial trees is even higher. Therefore, all the individuals in the initial M3GP population are created using the Full initialization method [6]. Additionally to the Full initialization, there was also an attempt to use deeper initial trees of depth 9 instead of 6. However, preliminary results did not show any improvement, and therefore the traditional initial depth of 6 levels was used.

Mutation. During the breeding phase, whenever mutation is the chosen genetic operator, one of three actions is performed, with equal probability: (1) standard subtree mutation, where a randomly created new tree replaces a randomly chosen branch (excluding the root node) of the parent tree; (2) adding a randomly created new tree as a new branch of the root node, effectively adding one dimension to the parent tree; and (3) randomly removing a complete branch of the root node, effectively removing one dimension from the parent tree.

As mentioned previously, M3GP begins with a population that only contains unidimensional individuals. From here, the algorithm has to be able to explore several different dimensions. In M3GP mutation is the only way of adding and removing dimensions, and therefore we have increased its probability of occurrence from 0.1 (used in M2GP [5]) to 0.5, to guarantee a proper search for the right dimension. Preliminary results have confirmed that a higher mutation rate indeed improves the fitness.

Crossover. Whenever crossover is chosen, one of two actions is performed, with equal probability: (1) standard subtree crossover, where a random node (excluding the root node) is chosen in each of the parents, and the respective branches swapped; (2) swapping of dimensions, where a random complete branch of the root node is chosen in each parent, and swapped between each other, effectively swapping dimensions between the parents. The second event is just a particular case of the first, where the crossing nodes are guaranteed to be directly connected to the root node.

Pruning. Mutation, as described above, makes it easy for M3GP to add dimensions to the solutions. However, many times some of the dimensions actually degrade the fitness of the individual, so they would be better removed. Mutation can also remove dimensions but, as described above, it does so randomly and blind to fitness. To maintain the simplicity and complete stochasticity of the genetic operators, we have decided not to make any of them more 'intelligent', and instead we remove the detrimental dimensions by pruning the best individual after the breeding phase.

The pruning procedure removes the first dimension and reevaluates the tree. If the fitness improves, the pruned tree replaces the original and goes through pruning of the next dimension. Otherwise, the pruned tree is discarded and the original tree goes through pruning of the next dimension. The procedure stops after pruning the last dimension.

Pruning is applied only to the best individual in each generation. Applying it to all the individuals in the population could pose two problems: (1) a significantly higher computational demand, where a considerable amount of effort would be spent on individuals that would still be unfit after pruning; (2) although not confirmed, the danger of causing premature convergence due to excessive removal of genetic material, the same way that code editing has shown to cause it [4].

Preliminary experiments have revealed that pruning the best individual of each generation shifts the distribution of the number of dimensions to lower values (or prevents it from shifting to higher values so easily) during the evolution, without harming fitness.

Elitism. It was mentioned earlier that, in order to explore solutions of different dimensions, M3GP relies on mutation to add and remove dimensions from the individuals, with a fairly high probability. It also has to rely on selection to keep the best dimensions in the population and discard the worst ones. The way to do this is by ensuring some elitism on the survival of the individuals from one generation to the next. M3GP does not allow the best individual of any generation to be lost, and always copies it to the next generation. Let us recall that this individual is already optimized in the sense that it went through pruning. Preliminary experiments have shown that elitism is indeed able to improve fitness.

5   Experimental Setup

This section describes the experiments performed to assess the performance of M3GP, in particular when compared to M2GP and other state-of-the-art classifiers.

Datasets. A set of eight problems was used for the experiments, the same used for M2GP [5]. This set contains both real world and synthetic data, having integer and real data types, with varying number of attributes, classes and samples. The 'heart' (HRT), 'segment' (SEG), 'vowel' (VOW), 'yeast' (YST) and 'movement-libras' (M-L) datasets can be found at the KEEL dataset repository [1], whereas the 'waveform' (WAV) dataset is available at [2]. 'IM-3' and 'IM-10' are the satellite datasets used in [11]. All the original datasets were randomly split in 70 % training and 30 % test sets, the same proportion as with M2GP [5]. Table 1 summarizes the main characteristics of each dataset.

Table 1. Data sets used for the experimental analysis.

Data Set	HRT	IM-3	WAV	SEG	IM-10	YST	VOW	M-L
No. of classes	2	3	3	7	10	10	11	15
No. of attributes	13	6	40	19	6	8	13	90
No. of samples	270	322	5000	2310	6798	1484	990	360

Tools. A modified version of GPLAB 3 was used to execute all the runs of M3GP. GPLAB is an open source GP toolbox for MATLAB, freely available at http://gplab.sourceforge.net. For the comparison with the state-of-the-art classifiers, we have used Weka 3.6.10. Weka is also open source, and freely available at http://www.cs.waikato.ac.nz/ml/weka/.

Parameters. Table 2 summarizes the parameters adopted for running M3GP. Some are the default parameters of GPLAB, unchanged from M2GP, while others have already been described in the previous section. In Weka we have used the default parameters and configurations for each algorithm.

6   Results and Discussion

This section presents comparative results between M2GP and M3GP, and also between M3GP and some of the best state-of-the-art classification methods used in machine learning.

Table 2. Running parameters of M3GP.

Runs	30
Population size	500 individuals
Generations	100 generations
Initialization	*6-depth Full initialization* [6]
Operator probabilities	Crossover $p_c = 0.5$, Mutation $p_\mu = 0.5$
Function set	$(+, -, \times, \div$ protected as in [6])
Terminal set	Ephemeral random constants $[0,1]$
Bloat control	17-depth limit [6]
Selection	Lexicographic tournament [9] of size 5
Elitism	Keep best individual

6.1 M2GP Versus M3GP

The comparison between M2GP and M3GP will be presented in terms of fitness, and in terms of number of nodes and number of dimensions of the solutions. Whenever a result is said to be significantly different (better or worse) from another, it means the difference is statistically significant according to the Wilcoxon's rank sum test for equal medians, performed at the 0.01 significance level.

Figures 2 and 3 show two sets of boxplots. Figure 2 reports the fitness obtained by the best individuals on each of the 30 training sets, while Fig. 3 reports the fitness obtained by these same individuals on the respective test sets. From now on we will call these the training fitness and the test fitness, respectively.

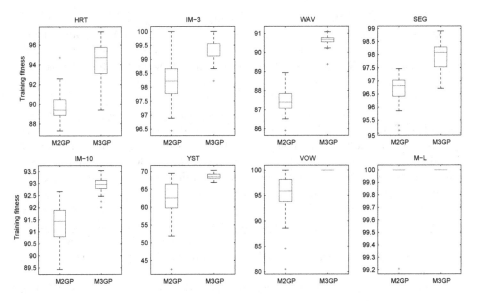

Fig. 2. Training fitness, given by classification accuracy, of M2GP and M3GP on all problems.

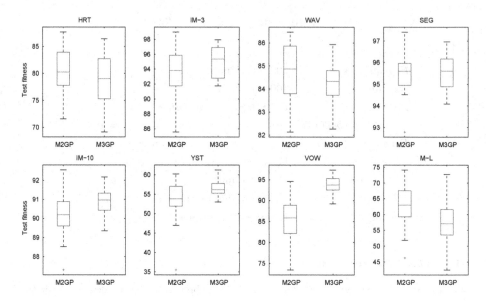

Fig. 3. Test fitness, given by classification accuracy, of M2GP and M3GP on all problems.

We may also call them training accuracy and test accuracy, respectively, since fitness is the accuracy of the classification. In each of these figures there is one boxplot for each problem. Each boxplot contains a pair of whiskered boxes, the first reporting the results of M2GP (most already reported in [5]) and the second reporting the results of M3GP.

It is clearly visible that M3GP achieves higher training fitness, which means it learns easier than M2GP, in all problems (in M-L the results of M2GP and M3GP are equal except for the outlier in M2GP). See Table 3 for numeric results and their statistical significance. M3GP is also able to achieve higher test fitness than M2GP in half of the problems. Once again, refer to Table 3 for the significance of these results.

Table 3 shows some quantitative results regarding the training and test fitness, also adding information on the number of nodes of the best individuals, as well as their number of dimensions. All these results refer to the median of the 30 runs. The best approach (between M2GP and M3GP) on each problem is marked in bold - both are marked when the difference is not statistically significant. In terms of size, we also consider lower to be better. However, we do not evaluate the number of dimensions qualitatively, since a higher number of dimensions does not necessarily translate into a larger number of nodes and/or lower interpretability of the solutions. We do include additional information for the number of dimensions, which is the minimum and maximum values obtained in the 30 runs.

Table 3 shows that, in terms of training fitness, M3GP is significantly better than M2GP in all the problems (except the last, M-L, where the results are

Table 3. Comparison between M2GP and M3GP.

	HRT	IM-3	WAV	SEG	IM-10	YST	VOW	M-L
Training fitness								
M2GP	89.4	98.2	87.4	96.8	91.4	62.6	95.9	100
M3GP	**94.7**	**99.6**	**90.7**	**98.1**	**93.0**	**68.5**	100	100
Test fitness								
M2GP	**80.2**	93.8	**84.9**	95.6	90.2	53.8	85.9	**63.0**
M3GP	79.0	**95.4**	84.3	95.6	**91.0**	**56.2**	**93.8**	57.1
Number of nodes								
M2GP	**37**	**24**	126	**43**	**117**	**146**	49	33
M3GP	110	66	**71**	111	239	274	**53**	**13**
Number of dimensions								
M2GP	2.5 *(1-8)*	2 *(1-4)*	5 *(2-10)*	4 *(3-8)*	7 *(4-10)*	5.5 *(1-13)*	9 *(4-18)*	10 *(7-12)*
M3GP	12 *(1-17)*	5 *(2-8)*	31 *(29-37)*	11 *(5-21)*	12 *(11-16)*	13 *(11-18)*	20 *(16-20)*	12 *(10-13)*

considered the same), while in terms of test fitness M3GP is better or equal to M2GP in all problems (except M-L). It is interesting to note that it is in the higher dimensional problems (except M-L) that M3GP achieves better results than M2GP (the problems are roughly ordered by dimensionality of the data). Problem M-L had already been identified as yielding a different behavior than the others [5], and here once again it is often the exception to the rule. Our explanation for M3GP not being able to perform better on this problem is the extreme easiness it has in reaching maximal accuracy. Both M2GP and M3GP achieve 100 % training accuracy, but M3GP does it in only a few generations (not shown), producing very small and accurate solutions that barely generalize to unseen data. On the other hand, M2GP does not converge immediately, so in its effort to learn the characteristics of the data it also evolves some generalization ability.

Regarding the size of the solutions, in most problems where M3GP brought improvements, it also brought significantly larger trees, except for WAV and M-L where the M3GP trees are significantly smaller, and VOW where the sizes are the same. However, when we split the nodes of the M3GP trees among their several dimensions, even the largest trees (e.g., in IM-10 and YST) seem to be simple and manageable (around 20 nodes per dimension), in particular when we consider that no simplification has been done except for the pruning of detrimental dimensions (see Sect. 4), and therefore the effective size of the trees may be even smaller.

Regarding the number of dimensions used in M2GP and M3GP, two things become clear. The first one is that there seems to be no single optimal number of dimensions for a given problem, since both M2GP and M3GP may choose wildly different values, depending on the run. The second one is that M3GP tends to use a larger number of dimensions than M2GP. What these numbers do not show is that different problems result in very different behaviors with respect to the evolution of the number of dimensions. Figure 4 illustrates two

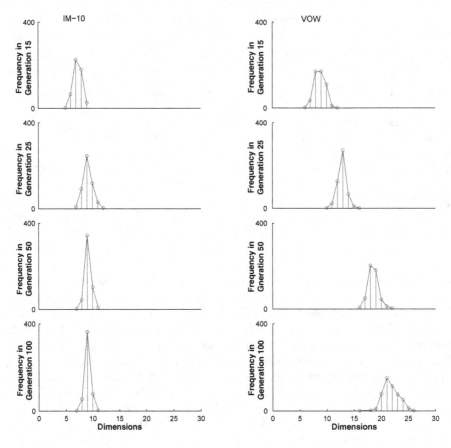

Fig. 4. Distribution of the number of dimensions in the population in generations 15, 25, 50 and 100 (top to bottom). On the left, a typical run of problem IM-10. On the right, a typical run of problem VOW.

main types of behavior, described next. In most problems the distribution of the number of dimensions moves rapidly to higher values in the beginning of the run, and then remains stable and more or less in the same range until the end of the run (exemplified on the left in Fig. 4). However, in some problems, like WAV and VOW, the distribution of the number of dimensions does not settle during the 100 generations of the run, and instead keeps moving towards higher values (exemplified on the right in Fig. 4). The WAV problem goes as high as 37 dimensions, and curiously this is one of the problems where M3GP produces significantly smaller trees than M2GP. VOW is another of the few problems where the M3GP trees are not larger than the M2GP trees. The only other such case is the unique M-L problem.

6.2   M3GP Versus State-of-the-art

The comparison between M3GP and the state-of-the-art classification methods is based only on training and test fitness. Based on the comparison previously done between M2GP and several state-of-the-art methods [5], we have decided to compare M3GP with a tree based classifier (RF - Random Forests), a meta classifier (RS - Random Subspace), and a function based classifier (MLP - Multi Layer Perceptron). The three of them were well ranked in the previous comparison with M2GP [5]. We have also included M2GP in this comparison to check how much better M3GP compares to the state-of-the-art than M2GP.

Table 4 reports and compares the training and test fitness obtained by RF, RS, MLP, M2GP and M3GP on the same eight problems, medians of 30 runs. The best approach on each problem is marked in bold, or several when their differences are not statistically significant. Looking at the first row, it is undeniable that RF is an almost unbeatable method when it comes to training fitness. Still, it is beaten by M3GP in the last two problems (VOW and M-L). (M2GP achieves the same feat in only one of them, M-L).

However, training fitness is not important unless accompanied by good test fitness, suggesting good generalization ability. Although RF is also good in test fitness, M3GP is able to achieve similar results. Like RF, M3GP is ranked first in five of the eight problems (M2GP achieves this is only two problems). Like RF, M3GP is not equaled by any other state-of-the-art method in two problems, WAV and VOW (M2GP achieves this only in WAV). We recall that these are precisely the two problems where the number of dimensions keeps growing during the entire evolution. We wonder if, given more generations, M3GP could distance itself even more from the other methods on these two problems. Regarding the other methods, MLP is ranked first in four problems, being the solo winner in one of them (M-L), while RS is ranked first in only two problems.

Table 4. Comparison between M3GP and state-of-the-art methods.

	HRT	IM-3	WAV	SEG	IM-10	YST	VOW	M-L
Training fitness								
RF	**98.4**	**100**	**99.5**	**99.9**	**99.8**	**98.3**	99.9	99.2
RS	88.9	97.1	92.0	98.4	96.3	71.1	97.8	92.3
MLP	**98.4**	98.7	98.5	97.6	91.0	64.6	91.9	91.3
M2GP	89.4	98.2	87.4	96.8	91.4	62.6	95.9	**100**
M3GP	94.7	99.6	90.7	98.1	93.0	68.5	**100**	**100**
Test fitness								
RF	**80.2**	94.8	81.5	**97.3**	**96.9**	**57.5**	89.4	71.8
RS	**81.5**	92.8	82.2	96.0	93.9	**56.6**	82.8	65.7
MLP	**80.2**	**95.9**	83.3	96.3	90.2	**58.0**	82.5	**75.9**
M2GP	**80.2**	93.8	**84.9**	95.6	90.2	53.8	85.9	63.0
M3GP	**79.0**	95.4	84.3	95.6	91.0	**56.2**	**93.8**	57.1

Besides the remarkable fact that M3GP achieves the same quality of results as the popular and successful RF in terms of test fitness, it is also worth remarking that the models provided by M3GP are potentially much easier to interpret than the ones provided by RF, or by any of the other two state-of-the-art methods.

7   Conclusions and Future Work

This paper addresses the problem of multiclass classification with GP, an area where previous approaches tended to yield poor performance. In particular, this paper presents M3GP, an extension of the recently proposed M2GP algorithm, a classifier that evolves transformations of the form $k(\mathbf{x}) : \mathbb{R}^p \to \mathbb{R}^d$, and applies a minimum Mahalanobis distance classifier. M3GP allows the search to consider a single dimension ($d = 1$) on which to transform the data at the beginning of the search, and progressively builds more dimensions guided by classifier performance.

The results are very encouraging. M3GP can deal with difficult benchmark and real-world problems and achieve state-of-the-art performance, comparing favorably with such methods as Random Forests, Random Subspaces and Multilayer Perceptron. Moreover, it is clear that M3GP adjusts its search based on the characteristics of each problem, automatically determining the best number of new feature dimensions to build in order to maximize accuracy.

Future work must consider a couple of limitations of the approach. First, M3GP needs to be fitted with some procedure to limit/prevent overfitting when accuracy on the training cases is easy to optimize (such as in the M-L problem). Another important aspect is to encourage the evolution of simple and small solutions, with the inclusion of bloat control or more efficient simplification strategies. Nonetheless, for now it is clear that M3GP is a general purpose and simple algorithm that is well worth pursuing and improving for use in challenging classification tasks.

Acknowledgments. This work was partially supported by FCT funds (Portugal) under contract UID/Multi/04046/2013 and projects PTDC/EEI-CTP/2975/2012 (MaSSGP), PTDC/DTP-FTO/1747/2012 (InteleGen) and EXPL/EMS-SIS/1954/2013 (CancerSys). Funding was also provided by CONACYT (Mexico) Basic Science Research Project No. 178323, DGEST (Mexico) Research Projects No. 5149.13-P and 5414.11-P, and FP7-Marie Curie-IRSES 2013 project ACoBSEC. Finally, the first author is supported by scholarship No. 372126 from CONACYT.

References

1. Alcala-Fdez, J., Fernandez, A., Luengo, J., Derrac, J., Garcia, S., Sanchez, L., Herrera, F.: Keel data-mining software tool: data set repository, integration of algorithms and experimental analysis framework. J. Multiple-Valued Log. Soft Comput. **17**(2–3), 255–287 (2011)
2. Bache, K., Lichman, M.: UCI Machine Learning Repository, University of California, Irvine, School of Information and Computer Sciences (2013). http://archive.ics.uci.edu/ml. Accessed 26 January 2015

3. Espejo, P.G., Ventura, S., Herrera, F.: A survey on the application of genetic programming to classification. Trans. Sys. Man Cyber Part C **40**(2), 121–144 (2010)

4. Haynes, T.: ollective adaptation: the exchange of coding segments. Evol. Comput. **6**(4), 311–338 (1998). http://dx.doi.org/10.1162/evco.1998.6.4.311

5. Ingalalli, V., Silva, S., Castelli, M., Vanneschi, L.: A multi-dimensional genetic programming approach for multi-class classification problems. In: Nicolau, M., et al. (eds.) 17th European Conference on Genetic Programming. LNCS, vol. 8599, pp. 48–60. Springer, Granada (2014)

6. Koza, J.R.: Genetic Programming: On the Programming of Computers by Means of Natural Selection, vol. 1. MIT press, Cambridge (1992)

7. Koza, J.R.: Human-competitive results produced by genetic programming. Genet. Program. Evol. Mach. **11**(3–4), 251–284 (2010)

8. Lin, J.Y., Ke, H.R., Chien, B.C., Yang, W.P.: Designing a classifier by a layered multi-population genetic programming approach. Pattern Recogn. **40**(8), 2211–2225 (2007)

9. Luke, S., Panait, L.: Lexicographic parsimony pressure. In: Proceedings of GECCO-2002, pp. 829–836. Morgan Kaufmann Publishers (2002)

10. Poli, R., Langdon, W.B., Mcphee, N.F.: A field guide to genetic programming. Lulu.com (Mar 2008)

11. U.S. Geological Survey (USGS): Earth resources observation systems (EROS) data center (EDC) (2015). http://glovis.usgs.gov/. Accessed 26 January 2015

12. Zhang, Y., Rockett, P.I.: A generic multi-dimensional feature extraction method using multiobjective genetic programming. Evol. Comput. **17**(1), 89–115 (2009)

Evolving Ensembles of Dispatching Rules Using Genetic Programming for Job Shop Scheduling

John Park[1][✉], Su Nguyen[1,2], Mengjie Zhang[1][✉], and Mark Johnston[1]

[1] Evolutionary Computation Research Group,
Victoria University of Wellington, PO Box 600, Wellington 6140, New Zealand
{John.Park,Su.Nguyen,Mengjie.Zhang}@ecs.vuw.ac.nz,
Mark.Johnston@msor.vuw.ac.nz
[2] International University - VNU HCMC, Ho Chi Minh City, Vietnam

Abstract. Job shop scheduling (JSS) problems are important optimisation problems that have been studied extensively in the literature due to their applicability and computational difficulty. This paper considers static JSS problems with makespan minimisation, which are NP-complete for more than two machines. Because finding optimal solutions can be difficult for large problem instances, many heuristic approaches have been proposed in the literature. However, designing effective heuristics for different JSS problem domains is difficult. As a result, hyper-heuristics (HHs) have been proposed as an approach to automating the design of heuristics. The evolved heuristics have mainly been priority based dispatching rules (DRs). To improve the robustness of evolved heuristics generated by HHs, this paper proposes a new approach where an ensemble of rules are evolved using Genetic Programming (GP) and cooperative coevolution, denoted as Ensemble Genetic Programming for Job Shop Scheduling (EGP-JSS). The results show that EGP-JSS generally produces more robust rules than the single rule GP.

Keywords: Genetic programming · Job shop scheduling · Hyper-heuristics · Ensemble learning · Cooperative coevolution · Robustness · Dispatching rules · Combinatorial optimisation · Evolutionary computation

1 Introduction

Job shop scheduling (JSS) problems are important optimisation problems that have been studied for over 50 years. JSS is still studied extensively due to its complexity and wide applications. JSS problems involve determining the optimal sequence to process jobs on the machines in a manufacturing system. For a JSS problem instance, each job has operations that need to be completed on different machines in a given sequence. However, a machine cannot process more than one job at a time. All operations must be processed by the machines to get a schedule, and the 'quality' of the solution generated for the JSS problem instance is given by the objective function. There are a number of existing approaches to solving JSS problems. Mathematical optimisation techniques give optimal solutions for

© Springer International Publishing Switzerland 2015
P. Machado et al. (Eds.): EuroGP 2015, LNCS 9025, pp. 92–104, 2015.
DOI: 10.1007/978-3-319-16501-1_8

static JSS problem instances. On the other hand, heuristic approaches, such as dispatching rules (DRs), have been applied to JSS to produce good solutions for large problem instances. Dispatching rules [12] are local decision makers which iteratively decide a sequence of jobs to be processed by a machine. In addition, meta-heuristic approaches [10,17] have also been applied to JSS. However, an issue with heuristic approaches to JSS is that they need to be carefully designed. Heuristic approaches also tend to be problem domain specific. Heuristics that are effective in one domain are not necessarily effective in other domains. Because of this, hyper-heuristics (HHs) [2] aim to automate the generation of heuristics such as DRs. However, DRs are limited as they make a single decision for choosing the next job to be processed by a machine. The myopic nature of DRs, combined with the fact that complex decisions need to be made for JSS problems, means that it is possible that DRs make bad decisions for certain situations within a particular JSS problem instance.

In classification, similar issues arise as single constituent rules cannot represent the noisy and complex decision boundaries between different classes sufficiently [13]. Because of this, ensemble approaches have been proposed [1,4], which have successfully been applied to difficult classification problems [13]. In an ensemble, a group of small constituent rules 'vote' on the outcomes. For example, the class labels represent the outcomes that can be 'voted' for in classification problem. It may be possible that ensembles of DRs can be used to deal with the complex decisions of selecting jobs better than single DRs, and improve the robustness of rules for JSS. However, ensemble approaches have not been seriously investigated for JSS.

The goal of this paper is to determine whether ensemble approaches can be used effectively for static JSS problem instances. An evaluation scheme is needed that allows a diverse set of rules to be evolved, as diversity is a cornerstone of ensemble approaches [13]. We denote this approach as Ensemble Genetic Programming for Job Shop Scheduling (EGP-JSS). This will be compared with an approach of evolving a single priority rule, denoted Genetic Programming for Job Shop Scheduling (GP-JSS). GP-JSS makes minor adjustments from a previous approach [11] of evolving DRs from GP by modifying the terminal set. Specific research objectives in this paper are:

(a) Developing a job selection procedure for the ensemble of rules for JSS.
(b) Developing a new fitness function for EGP-JSS to ensure that a diverse set of rules are evolved.
(c) Comparing the evolved ensemble rules by EGP-JSS and GP-JSS with the benchmark DRs.

2 Background

This section briefly describes some background on the JSS problem with previous approaches for JSS, and the hyper-heuristic approaches that have been applied to JSS.

2.1 Job Shop Scheduling Problem

A JSS problem instance consists of N jobs and M machines, and a list of operations for each job. Compared to dynamic JSS problems, static JSS problems have all attributes of jobs, machines and operations known from the beginning, and do not contain any stochastic elements. An operation σ_{ij} in a JSS problem instance is the ith operation of job j, and $M(\sigma_{ij})$ denotes the machine that the operation is processed on. An operation σ_{ij} can only be carried out when operation σ_{i-1j} has been completed (with σ_{1j} being the first operation of a job j), and when the machine to be processed on $(M(\sigma_{ij}))$ is available. The time when a machine i is available is denoted as R_{M_i}. Each job j has a ready time $r(\sigma_{1j})$ for when its first operation is available, and each operation σ has processing time $p(\sigma)$, and setup time $s(\sigma)$. The number of operations for job j is N_j, and the total remaining processing time is $\sum_{k=i}^{N_j} p(\sigma_{kj})$. For this paper, we focus on the static JSS problem with makespan minimisation, i.e., minimising the maximum completion time $C_{\max}$. This is denoted as $Jm||C_{\max}$.

$Jm||C_{\max}$ for $M = 2$ machines can be solved optimally via Jackson's algorithm [12]. However, Garey et al. [5] showed that the JSS makespan minimisation problem is NP-complete for $M > 2$. In JSS problems with instances that have hundreds of jobs and a large number of machines [15], exact optimisation is too computationally expensive. For such JSS problem instances, the primary approaches use heuristics, such as DRs [12] and meta-heuristics. DRs range in complexity from basic first-in-first-out (FIFO) rules, which processes the jobs in the order they arrive, to more complex composite dispatching rules (CDRs) [9], which combine smaller heuristics to form custom made priority functions. On the other hand, a wide range of meta-heuristic approaches have been proposed in the literature. Meta-heuristic approaches include Simulated Annealing [10] and Genetic Algorithms (GA) [17].

2.2 Genetic Programming Based Hyper-Heuristic Approaches

In conjunction with heuristic and meta-heuristic approaches, hyper-heuristics (HHs) [2] have also been investigated for JSS. Instead of searching the solution space directly, HHs are given heuristic components to generate heuristics with, and a fitness measure to evaluate how well generated heuristics perform. It then searches for a good heuristic, optimising over the fitness measure. A number of HH approaches to JSS in the literature use Genetic Programming (GP) [2].

Dimopoulos and Zalzala [3] use GP to evolve priority based DRs for a single machine JSS problem. An arithmetic representation consisting of mathematical operators and job attributes are used to represent the individuals in the GP system. They showed that the evolved rules performed better than the man-made benchmark DRs. Geiger et al. [6] use GP to evolve priority based DRs for various single machine JSS problems in both static and dynamic environments. They showed that GP can evolve DRs that can generate optimal solutions for some special static single machine JSS problems with polynomial time exact algorithms, and evolve effective rules for NP-hard JSS problems.

Jakobovic et al. [8] proposed a GP based hyper-heuristic approach to evolving priority based DRs for the multi-machine static and dynamic JSS problems ranging from 3 to 20 machines. Tay and Ho [16] proposed a priority based GP approach to multi-objective flexible job-shop problems, and showed that the evolved rules outperformed other simple DRs. However, later examination [7] showed that Tay and Ho's approach [16] does not perform as well in different dynamic job shop scenarios. Nguyen et al. [11] compared three different representations for GP to evolve DRs for static JSS problems. The first representation they propose is a decision tree representation (R_1), where the individuals are given DRs and make decisions on which rule to use for dispatch jobs onto available machines. The second representation is an arithmetic representation (R_2) where the individuals represent priority function trees. The third representation (R_3) combines both R_1 and R_2 representations, where an individual can define its own priority function tree that is used in conjunction with the decision tree. They showed that out of the three GP representations, R_3 performed better than both R_1 and R_2. In addition, they showed that the evolved rules are competitive with meta-heuristics such as a hybrid GA [17] proposed in the literature.

3   The New Approaches

This section proposes two approaches. The first approach evolves simple priority based dispatching rules denoted Genetic Programming for Job Shop Scheduling (GP-JSS) approach. This extends Nguyen et al.'s [11] arithmetic representation for GP to evolve dispatching rules, and will be used as a benchmark. The second approach is EGP-JSS, which evolves an ensemble of priority rules simultaneously.

3.1   GP Representation

For both GP-JSS and EGP-JSS, the dispatching rules generated are non-delay. In a non-delay schedule, a job is selected to be processed on machine i as soon as machine i is ready to process a new job if there are any jobs waiting to be processed at that machine. We denote the number of idle jobs waiting at a machine i as W_i. Tree-based GP is used, and the individuals in the GP population represent arithmetic function trees. The function trees generate priorities for the jobs waiting to be processed by machine i. How these priorities are used to select the job to process differs between GP-JSS and EGP-JSS, and is discussed further below.

The terminal set consists of the properties of the job shop scheduling environment discussed in Sect. 2.1. These are shown in Table 1. These extend the terminal set used by Nguyen et al. [11] in their comparison of different GP representations. The new added terminals are the number of waiting jobs (NJ), and a sufficiently large value (LV). The function set consists of the operators $+$, $-$, $\times$, protected division $/$, and if. For the ternary if operator, the value of the second subtree if will be returned if the value of the first subtree representing the conditional is ≥ 0; otherwise, the value of the third subtree else is returned.

Table 1. The terminal set used for the GP representations, where job j is one of the job waiting to be processed as soon as machine i is ready.

Terminal	Description	Value
RJ	Operation ready time	$r(\sigma_{ji})$
RO	Remaining number of operations of job j	$N_j - i + 1$
RT	Remaining total processing times of job j	$\sum_{k=i}^{N_j} p(\sigma_{jk})$
PR	Operation processing time	$p(\sigma_{ji})$
RM	Machine ready time	R_{M_i}
NJ	Idle jobs waiting at machine	W_i
#	Constant	Uniform[0,1]
LV	Sufficiently large value	∞

To evaluate an individual x in the GP population, the individual is used as a non-delay dispatching rule to generate solutions on T_{train} sample training JSS problem instances. For each JSS problem instance I, a lower bound LB$_I$ is calculated for the makespan as specified by Taillard [15]. From the solution, the makespan objective, $Obj(x, I)$, is calculated and the deviation dev_I of $Obj(x, I)$ from LB$_I$, as shown in Eq. (1), is used as the fitness value for individual x for the specific problem instance I. The average fitness $fitness_{avg}(x)$ of individual x over the entire training set T_{train} is given by Eq. (2).

$$fitness(x, I) = dev_I = \frac{Obj(x, I) - \text{LB}_I}{\text{LB}_I} \tag{1}$$

$$fitness_{avg}(x) = \frac{1}{T_{train}} \sum_{t=1}^{T_{train}} fitness(x, I_t) \tag{2}$$

For the EGP-JSS approach, we use two fitness functions. The first fitness function is simply the one used for GP-JSS, where $fitness(x) = fitness_{avg}(x)$. This is denoted as 'No Fitness Modification' (NFM). The second fitness function takes diversity of the indviduals in the ensemble into account by penalising similar individuals, and is denoted as 'With Fitness Modification' (WFM). WFM is covered in detail in Sect. 3.3.

3.2 Genetic Programming for Job Shop Scheduling (GP-JSS)

The GP-JSS approach uses a GP population of individuals to evolve a single tree as its output. GP-JSS is an extension of the R_2 representation proposed by Nguyen et al. [11] that uses the extended set of terminals provided in Table 1. How a job is selected in a non-delay priority based DR is illustrated in Fig. 1. When selecting which job to process for a free machine, an individual in the population is used to assign priority values to each of the idle jobs waiting to be processed by the machine. The job with the highest priority is then selected to be processed. This continues until all operations have been completed.

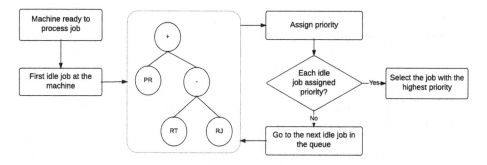

Fig. 1. Priority based dispatching rule job selection for available machine.

3.3 Ensemble Genetic Programming for Job Shop Scheduling (EGP-JSS)

EGP-JSS evolves dispatching rules which are used in an ensemble of priority rules to determine which job to process for a ready machine. However, using a single population for ensembles will require a carefully designed grouping scheme to group the individuals together, along with a complementary evaluation scheme to evaluate the grouped individuals. Instead of doing this, we consider an approach where we partition the population into S smaller subpopulations. Each subpopulation has size K. EGP-JSS groups the individuals from the different subpopulations together to form an ensemble. This approach of splitting the population into smaller subpopulations that work together to solve a problem is known as cooperative coevolution [14]. By using cooperative coevolution, we allow for the subcomponents of the ensemble to apply crossover, mutation and reproduction separately, and allow for diversity between the different subcomponents.

In cooperative coevolution, individuals in a subpopulation only interact with representatives of the other subpopulations when they are being evaluated for their fitness. A representative is defined as the individual with the best fitness in a subpopulation. Initially, before the first fitness evaluation, the representative of each subpopulation is chosen randomly. Unlike Potter and De Jong's [14] cooperative coevolution approach, we do not destroy unproductive subpopulations, as destroying and regenerating a new subpopulation of individuals will require a large number of generations for it to be effective.

The pseudocode of the EGP-JSS approach is shown in Algorithm 1. The job selection procedure and the fitness evaluation scheme is discussed further below.

Job Selection Procedure. As shown in Fig. 2, for rules evolved using EGP-JSS, the decision of choosing a job for a ready machine is carried out by the individual from the different subpopulations 'voting' on the jobs, and taking the job with the most votes. An individual 'votes' for the job if the job has the highest priority assigned to it by the individual. An individual's 'voting' procedure works similar to the job selection procedure for priority based DR described for GP-JSS.

Data: S, K, T_{train}, number of generations G, fitness evaluation scheme *eval*
Result: Representative individuals $x'_1, \ldots, x'_S$
Initialise GP subpopulations $\nabla_1, \ldots, \nabla_S$
for *each subpopulation* ∇_1 **to** ∇_S **do**
   | $x'_i \leftarrow$ random individual from ∇_i
end
while G *number of generations has not yet passed* **do**
   for *each subpopulation* ∇_1 **to** ∇_S **do**
      for *each individual* x *in* ∇_i **do**
         form an ensemble $E = \{x, x'_1, \ldots, x'_S\} - \{x'_i\}$
         for *each instance* I *in training* T_{train} **do**
            `/* solve I using E as a non-delay dispatching rule */`
            while *leftover operations remaining* **do**
               if *machine* i *is available* **then**
                  $j \leftarrow selection(E, j_1, \ldots, j_{W_i})$
                  process job j on machine i
               end
            end
            $fitness(x, I) \leftarrow$ fitness of solution
         end
         `/* eval denotes the fitness evaluation scheme */`
         $fitness(x) \leftarrow eval(fitness(x, I_1), \ldots, fitness(x, I_{T_{train}}))$
         update x'_i if $fitness(x) > fitness(x'_i)$
      end
   end
end

Algorithm 1. The pseudocode for the EGP-JSS approach.

If there is a tie in the votes, e.g., two jobs, j_1 and j_2 have the same number of votes as each other, a tie-breaker scheme is carried out. For an individual rule x, let $\delta_x(j_1), \ldots, \delta_x(j_{W_i})$ be the priorities assigned to jobs $j_1, \ldots, j_{W_i}$ waiting to be processed at a machine. The normalised priority of a job j, is defined by Eq. (3), where $f(j) = \frac{1}{1+e^{-\delta_x(j)}}$.

$$\delta'_x(j) = \frac{f(j)}{\sum_{r=1}^{W_i} f(j_r)} \tag{3}$$

Afterward, the job with the highest sum of priority values over all ensemble members out of the top voted jobs is then selected for processing.

With Fitness Modification (WFM) Evaluation Scheme. The WFM fitness function, which takes diversity of individuals into account, is defined as follows. To evaluate the diversity of an individual, the phenotype of individuals in a subpopulation are compared against the representative individuals of the other subpopulations. In this case, the phenotype is defined as the list (of length L_I) of all the priorities that are calculated for the jobs as the solution for

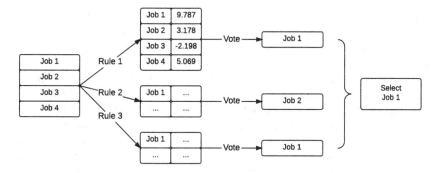

Fig. 2. Example of ensemble job selection process for an available machine.

the problem instance I is being generated. This includes the priorities that are calculated for waiting jobs that were not selected for processing at a particular moment of decision. These are normalised on the interval $[0, 1]$ using a sigmoid function $g(x, z) = \frac{1}{1+e^{-\delta_x(z)}}$, where x is the GP individual being evaluated, and $\delta_x(z)$ is the zth priority calculated by x.

After all the priorities are normalised, the penalty is the average of the squared differences between the priorities of the individual x of a subpopulation to the representative individuals y of the other subpopulations, as shown by Eq. (4).

$$penalty(x, I) = 1 - \sum_{y=1, y \neq x}^{S} \sum_{z=1}^{L_I} \frac{(g(x, z) - g(y, z))^2}{(S-1)L_I} \tag{4}$$

To incorporate the penalty into the fitness evaluation of an individual x in the subpopulation, the average penalty of the individual $penalty_{avg}(x)$ is calculated over all problem instances by taking the mean of penalties. The average fitness from Eq. (2) is then multiplied by one plus the average penalty to get the final fitness $fitness(x) = fitness_{avg}(x)(1 + penalty_{avg}(x))$. This means that when an individual x from a subpopulation is very different from the representatives of the other subpopulations, the $penalty_{avg}(x) \approx 0$, and hence $fitness(x) \approx fitness_{avg}(x)$.

4 Experimental Design

For training and testing, we use the JSS benchmark dataset proposed by Taillard [15]. The dataset consists of 8 sets of 10 problem instances broken up by the number of jobs and the number of machines. All jobs in each problem instance have zero release times and setup times, and must be processed on all machines.

For training, we use three separate sets of JSS problem instances from the Taillard dataset. The first training set Δ_1 is the first five problem instances from the set of data containing $N = 15$ jobs and $M = 15$ machines. The second training set Δ_2 is the first five problem instances from the set of data containing $N = 30$ jobs and $M = 20$ machines. The third training set Δ_3 is the first five problem instances from the set of data containing $N = 100$ jobs and $M = 20$ machines. The standard GP-JSS approach has population size of 1024. For the EGP-JSS approach, given a fixed number of subpopulations S, the subpopulation size is given by $K = \lfloor \frac{1024}{S} \rfloor$. This gives us a total number of individuals in the EGP-JSS approach that is approximately equal to the population size of the GP-JSS approach. For $S = 3, 4, 5, 6, 7, 8, 9, 10$, this gives us $K = 341, 256, 204, 170, 146, 128, 113, 102$ respectively. These are shown in Table 2 with the notation $\langle S, K \rangle$, along with the other parameters used for GP. The GP-JSS and the EGP-JSS approaches were run over each training set 30 times using different seeds, resulting in 30 evolved dispatching rules over each training set. For testing, the problem instances that are not used in the training sets Δ_1, Δ_2 or Δ_3 are used, meaning that there are 65 problem instances in the test set.

Table 2. GP parameters used for evolving rules

Parameter	GP-JSS Value	EGP-JSS Value
$\langle$Subpopulations, Subpopulation sizes$\rangle$	$\langle 1, 1024 \rangle$	$\langle 3, 341 \rangle$, $\langle 4, 256 \rangle$, $\langle 5, 204 \rangle$, $\langle 6, 170 \rangle$, $\langle 7, 146 \rangle$, $\langle 8, 128 \rangle$, $\langle 9, 113 \rangle$, $\langle 10, 102 \rangle$
Crossover rate	80 %	80 %
Mutation rate	10 %	10 %
Reproduction rate	10 %	10 %
Generations	51	51
Max-depth	8	8
Selection method	tournament selection	tournament selection
Selection size	7	7
Initialisation	ramped-half-and-half	ramped-half-and-half

The R_2 representation proposed by Nguyen et al. [11] is used as a benchmark for GP-JSS and EGP-JSS. R_2 will have the same parameter settings as the GP-JSS approach. As a benchmark, the R_2 representation proposed by Nguyen et al. [11] is used to compare the robustness of the rules evolved using GP-JSS and EGP-JSS. Afterward, the GP-JSS and the EGP-JSS approaches are compared against benchmark DRs. The first two benchmarks are simple non-delay schedules that select jobs to process on an available machine by the order of their arrival (FIFO); and selecting jobs by the shortest processing time (SPT). The other benchmarks are the best rules evolved by Nguyen et al. [11] for their R_1, R_2 and R_3 representations, which are used as non-delay dispatching rules. As evolved rules from EGP-JSS are not improvement heuristics, they are not compared against state-of-the-art meta-heuristic approaches to static JSS which compensate for long running time by producing very good solutions to static JSS problem instances.

5   Results

The solution's deviation dev_I (see Eq. (1)) from the lower bound is used for measuring the quality of the solution generated by the DRs. Afterward, the average of all the problem instances, denoted as dev_{avg}, is used for evaluating the DRs over the entire test set. In the tables that follow, sets of rules evolved by EGP-JSS that perform significantly better than the rules evolved by GP-JSS and R_2 are marked with †. The standard z-test is used to compare the DRs against each other. One set of evolved rules is considered significantly better than another if the obtained p-value under the statistical test is less than 0.05.

5.1   Parameter Settings Evaluation

First, the different ⟨Subpopulations, Subpopulation sizes⟩ in Table 2 are compared against each other to find the 'best' configuration. In addition, we compare WFM and NFM against each other to see whether the modified evaluation scheme (WFM) improves the performance of the evolved rules. The preferred configuration for EGP-JSS is used for comparison against the other benchmarks. This is shown in Table 3.

From the results of Table 3, we can see that the results of EGP-JSS under different parameter settings are similar to each other. No configuration is significantly better than other configurations. This means that when K scales with

Table 3. dev_{avg} of evolved rules from EGP-JSS for the Taillard's dataset for different ⟨S, K⟩ and for the fitness functions WFM and NFM

		Δ_1	Δ_2	Δ_3	Testing
WFM	⟨3, 341⟩	0.47 ± 0.07	0.36 ± 0.08	0.06 ± 0.05	0.28 ± 0.06
	⟨4, 256⟩	0.45 ± 0.03	0.33 ± 0.04	0.05 ± 0.02	0.26 ± 0.03
	⟨5, 204⟩	0.45 ± 0.04	0.34 ± 0.03	0.05 ± 0.02	0.26 ± 0.02
	⟨6, 170⟩	0.47 ± 0.04	0.35 ± 0.05	0.06 ± 0.04	0.27 ± 0.04
	⟨7, 146⟩	0.46 ± 0.04	0.34 ± 0.04	0.06 ± 0.03	0.27 ± 0.03
	⟨8, 128⟩	0.47 ± 0.05	0.35 ± 0.04	0.06 ± 0.03	0.27 ± 0.03
	⟨9, 113⟩	0.46 ± 0.03	0.35 ± 0.04	0.06 ± 0.03	0.27 ± 0.03
	⟨10, 102⟩	0.46 ± 0.03	0.35 ± 0.04	0.06 ± 0.02	0.27 ± 0.02
NFM	⟨3, 341⟩	0.49 ± 0.08	0.36 ± 0.08	0.07 ± 0.06	0.29 ± 0.07
	⟨4, 256⟩	0.47 ± 0.07	0.35 ± 0.07	0.06 ± 0.05	0.27 ± 0.06
	⟨5, 204⟩	0.46 ± 0.05	0.35 ± 0.05	0.06 ± 0.04	0.27 ± 0.05
	⟨6, 170⟩	0.46 ± 0.05	0.35 ± 0.05	0.06 ± 0.03	0.27 ± 0.04
	⟨7, 146⟩	0.45 ± 0.02	0.34 ± 0.03	0.05 ± 0.02	0.26 ± 0.02
	⟨8, 128⟩	0.45 ± 0.04	0.34 ± 0.04	0.05 ± 0.03	0.26 ± 0.03
	⟨9, 113⟩	0.45 ± 0.02	0.34 ± 0.02	0.05 ± 0.01	0.26 ± 0.01
	⟨10, 102⟩	0.46 ± 0.03	0.34 ± 0.03	0.05 ± 0.02	0.26 ± 0.02

S, the value of S is not significant to the performance of the evolved rules under the EGP-JSS approach.

5.2   GP-JSS and EGP-JSS

From the results of Sect. 5.1, we selected the configuration with $\langle 4, 256 \rangle$ that uses the modified fitness measure WFM to be compared against the GP-JSS and R_2 approaches. Although $\langle 4, 256 \rangle$ with WFM is not significantly better than the other configurations, it had the lowest mean deviation for the test set. For each approach, 30 rules are evolved using the training sets Δ_1, Δ_2 and Δ_3, and their performances over training runs and the test runs over the respective training and the test sets are used. This is shown in Table 4.

Table 4. dev_{avg} of evolved rules from GP-JSS and EGP-JSS for $Jm||C_{\max}$.

	Training			Testing		
	R_2	GP-JSS	EGP-JSS	R_2	GP-JSS	EGP-JSS
Δ_1	0.59 ± 0.15	0.57 ± 0.11	$0.45 \pm 0.03^\dagger$	0.37 ± 0.13	0.36 ± 0.12	$0.26 \pm 0.04^\dagger$
Δ_2	0.40 ± 0.15	0.40 ± 0.11	$0.33 \pm 0.04^\dagger$	0.32 ± 0.13	0.32 ± 0.10	$0.26 \pm 0.03^\dagger$
Δ_3	0.11 ± 0.10	0.12 ± 0.10	$0.06 \pm 0.01^\dagger$	0.32 ± 0.13	0.34 ± 0.12	$0.26 \pm 0.01^\dagger$

Although the GP-JSS extended the R_2 approach by adding more terminals to the terminal set, we can see in Table 4 that it did not improve on the original approach significantly. It is likely that the added terminals representing the number of idle jobs waiting at the machine (NJ) and sufficiently large value (LV) are not important to the sequencing decisions that are made by the DRs.

However, we can see that the rules evolved using EGP-JSS perform significantly better than the rules evolved using the GP-JSS and R_2 approaches, outperforming the other rules evolved under the three training sets Δ_1, Δ_2 and Δ_3. In addition, the rules evolved with EGP-JSS have much lower standard deviations, meaning that the evolved rules mostly performed similar to each other and are more stable than those evolved with GP-JSS and R_2. The results show that EGP-JSS can potentially produce more robust rules than the "standard" approach.

5.3   Evolved Rules and Benchmark Dispatching Rules

The final evaluation compares the best rules evolved from each training set using GP-JSS and EGP-JSS against other dispatching rules over the training and the test sets. The best evolved rules from GP-JSS and EGP-JSS are denoted as Θ_1^{GP} and Θ_1^{EGP} respectively, where the subscript on Θ denotes each training set (e.g. Θ_1 means best rule trained over Δ_1). The first two benchmarks are non-delay FIFO and SPT dispatching rules. The three other benchmarks are rules evolved by Nguyen et al. [11] using the three different representations for individuals in the GP population. The best rule from their R_1 representation, R_2 and R_3 are denoted as $\Theta_{R_1}^{c1}$, $\Theta_{R_2}^{c2}$ and $\Theta_{R_3}^{c3}$ respectively. This is shown in Table 5.

Table 5. Deviation of the DRs against the lower bound for the training sets (Δ_1, Δ_2, Δ_3) and the entire dataset.

Rule	Δ_1			Δ_2			Δ_3			Testing		
	Min	Avg	Max	Min	Avg	Max	Min	Avg	Max	Min	Avg	Max
FIFO	0.53	0.65	0.81	0.49	0.55	0.58	0.14	0.20	0.23	0.17	0.44	0.94
SPT	0.44	0.58	0.78	0.33	0.50	0.64	0.13	0.15	0.18	0.08	0.39	0.81
$\Theta_{R_1}^{c1}$	0.49	0.75	0.99	0.63	0.69	0.71	0.31	0.36	0.43	0.34	0.61	1.27
$\Theta_{R_2}^{c2}$	0.38	0.44	0.47	0.27	0.33	0.36	0.01	0.04	0.07	0.00	0.24	0.67
$\Theta_{R_3}^{c3}$	0.38	0.48	0.59	0.35	0.40	0.45	0.06	0.09	0.12	0.06	0.29	0.68
Θ_1^{GP}	0.35	0.44	0.56	0.27	0.31	0.34	0.03	0.05	0.09	0.01	0.24	0.63
Θ_2^{GP}	0.33	0.42	0.51	0.26	0.32	0.36	0.03	0.05	0.07	0.02	0.24	0.60
Θ_3^{GP}	0.38	0.44	0.47	0.28	0.31	0.36	0.01	0.03	0.07	0.02	0.25	0.57
Θ_1^{EGP}	0.37	0.43	0.47	0.30	0.32	0.37	0.03	0.05	0.08	0.01	0.24	0.58
Θ_2^{EGP}	0.38	0.44	0.47	0.24	0.31	0.36	0.01	0.04	0.07	0.00	0.24	0.67
Θ_3^{EGP}	0.38	0.44	0.47	0.28	0.33	0.36	0.01	0.04	0.07	0.03	0.24	0.62

From the results of Table 5, we can see that the best rules from the GP-JSS and the EGP-JSS approaches perform significantly better than the two simple DRs. This reinforces the idea that evolved rules outperform the simple DRs for JSS problems literature [3, 6, 11]. On the other hand, the best rules for GP-JSS and EGP-JSS perform similarly to $\Theta_{R_1}^{c1}$, $\Theta_{R_2}^{c2}$ and $\Theta_{R_3}^{c3}$.

6   Conclusions

In this paper, we proposed a novel approach (EGP-JSS) of evolving an ensemble of DRs using GP and cooperative coevolution. The experimental results show that the ensemble of rules evolved from the EGP-JSS approach perform significantly better than the benchmark GP-JSS and R_2 approaches. Including the two new terminals in GP-JSS does not significantly improve the performance over R_2. The rules evolved by EGP-JSS are more robust than the simple conventional rules FIFO and SPT.

For future work, extending the ensemble approach to dynamic JSS problem would be very interesting. In dynamic JSS problems properties of jobs are not known before they arrive at the shop floor. Because of this, global optimisation techniques used in static JSS do not work in dynamic JSS. Good robust dispatching rule approach will be required to handle the uncertainity in conjunction with the standard sequencing decisions in dynamic JSS. In addition, developing a GP based ensemble approach that uses a single population would also be very useful, as it removes the need to define the number of subpopulations and their respective sizes.

References

1. Breiman, L.: Bagging predictors. Mach. Learn. **24**(2), 123–140 (1996)
2. Burke, E.K., Gendreau, M., Hyde, M., Kendall, G., Ochoa, G., Ozcan, E., Qu, R.: Hyper-heuristics: a survey of the state of the art. J. Oper. Res. Soc. **64**(12), 1695–1724 (2013)
3. Dimopoulos, C., Zalzala, A.M.S.: Investigating the use of genetic programming for a classic one-machine scheduling problem. Adv. Eng. Softw. **32**(6), 489–498 (2001)
4. Freund, Y., Schapire, R.: A decision-theoretic generalization of on-line learning and an application to boosting. In: Vitányi, P. (ed.) Computational Learning Theory. Lecture Notes in Computer Science, pp. 23–37. Springer, Berlin (1995)
5. Garey, M.R., Johnson, D.S., Sethi, R.: The complexity of flowshop and jobshop scheduling. Math. Oper. Res. **1**(2), 117–129 (1976)
6. Geiger, C.D., Uzsoy, R., Aytu, H.: Rapid modeling and discovery of priority dispatching rules: an autonomous learning approach. J. Sched. **9**(1), 7–34 (2006)
7. Hildebrandt, T., Heger, J., Scholz-Reiter, B.: Towards improved dispatching rules for complex shop floor scenarios: a genetic programming approach. In: Proceedings of the 12th Annual Conference on Genetic and Evolutionary Computation, pp. 257–264 (2010)
8. Jakobovi, D., Jelenkovi, L., Budin, L.: Genetic programming heuristics for multiple machine scheduling. In: Ebner, M., O'Neill, M., Ekárt, A., Vanneschi, L., Esparcia-Alcázar, A.I. (eds.) Genetic Programming. Lecture Notes in Computer Science, vol. 4445, pp. 321–330. Springer, Heidelberg (2007)
9. Jayamohan, M.S., Rajendran, C.: New dispatching rules for shop scheduling: a step forward. Int. J. Prod. Res. **38**(3), 563–586 (2000)
10. Kreipl, S.: A large step random walk for minimizing total weighted tardiness in a job shop. J. Sched. **3**(3), 125–138 (2000)
11. Nguyen, S., Zhang, M., Johnston, M., Tan, K.C.: A computational study of representations in genetic programming to evolve dispatching rules for the job shop scheduling problem. IEEE Trans. Evol. Comput. **17**(5), 621–639 (2013)
12. Pinedo, M.L.: Scheduling: theory, algorithms and systems development. In: Gaul, W., Bachem, A., Habenicht, W., Runge, W., Stahl, W.W. (eds.) Operations Research Proceedings 1991. Operations Research Proceedings 1991, vol. 1991, 3rd edn, pp. 35–42. Springer, Heidelberg (2012)
13. Polikar, R.: Ensemble based systems in decision making. IEEE Circuits Syst. Mag. **6**(3), 21–45 (2006)
14. Potter, M.A., De Jong, K.A.: Cooperative coevolution: an architecture for evolving coadapted subcomponents. Evol. Comput. **8**(1), 1–29 (2000)
15. Taillard, E.: Benchmarks for basic scheduling problems. Eur. J. Oper. Res. **64**(2), 278–285 (1993)
16. Tay, J.C., Ho, N.B.: Evolving dispatching rules using genetic programming for solving multi-objective flexible job-shop problems. Comput. Ind. Eng. **54**(3), 453–473 (2008)
17. Zhou, H., Cheung, W., Leung, L.C.: Minimizing weighted tardiness of job-shop scheduling using a hybrid genetic algorithm. Eur. J. Oper. Res. **194**(3), 637–649 (2009)

Attributed Grammatical Evolution Using Shared Memory Spaces and Dynamically Typed Semantic Function Specification

James Vincent Patten[(✉)] and Conor Ryan

Biocomputing and Developmental Systems Group,
University of Limerick, Limerick, Ireland
{james.patten,conor.ryan}@ul.ie

Abstract. In this paper we introduce a new Grammatical Evolution (GE) system designed to support the specification of problem semantics in the form of attribute grammars (AG). We discuss the motivations behind our system design, from its use of shared memory spaces for attribute storage to the use of a dynamically type programming language, Python, to specify grammar semantics.

After a brief analysis of some of the existing GE AG system we outline two sets of experiments carried out on four symbolic regression type (SR) problems. The first set using a context free grammar (CFG) and second using an AG. After presenting the results of our experiments we highlight some of the potential areas for future performance improvements, using the new functionality that access to Python interpreter and storage of attributes in shared memory space provides.

Keywords: Grammatical Evolution · Symbolic regression · Attribute grammars

1 Introduction

Since it was first introduced [6], Grammatical Evolution (GE) has been successfully applied to solve a wide range of problems across a diverse set of domains. GE operates by producing potential solutions (usually in the form of programs), to a predefined problem, by combining symbols specified in Backus-Naur Form (BNF), a convenient way of describing a Context Free Grammar (CFG).

A CFG provides a means of specifying the syntax of programs, by outlining a set of rules which control the sequences of symbols allowed to appear in each program. While a CFG provides a means of specifying program syntax, it does not support specification of semantics, information which could guide the generation of more meaningful programs.

A GE system uses the rules of a CFG specification in combination with an individuals genotype to produce the individuals phenotype. After a phenotype is successfully produced we can extract the parse tree from it and then use this parse tree to evaluate the fitness of the individual across a set of training data

© Springer International Publishing Switzerland 2015
P. Machado et al. (Eds.): EuroGP 2015, LNCS 9025, pp. 105–112, 2015.
DOI: 10.1007/978-3-319-16501-1_9

points. Usually in GE it is not until the assignment of fitness that issues of semantic correctness become apparent. A common practice is to include some means of detecting semantically invalid programs when running fitness evaluation, e.g. protected division, or assignment of worst fitness score to individuals whose fitness evaluation "throws" an error.

As fitness scores are used to decide which individuals get to act as parents during evolution and to decide which individuals to replace in a steady state population, the score assigned to an individual is very important. While semantically invalid individuals do "die out" due to the evolutionary process the effects of their initial introduction into a population is something that needs to be considered [4]. Also as training data sets become much larger and fitness evaluation time increases we need to more carefully consider the effects of evaluation time spent on individuals that eventually get assigned a worst fitness score. One method that has the potential to reduce these effects is the addition of semantic information to help guide the genotype to phenotype mapping process, ensuring individuals produced are not only syntactically but semantically correct.

Knuth [5] proposed a means of annotating a CFG with semantic information in the form of attributes and semantic functions, commonly referred to as Attribute Grammar (AG). Unlike a CFG, when used with GE in the creation of a derivation tree, an AG in addition to providing a set of production rules, will also provide an associated semantic function which specify attributes to annotate the nodes of the derivation tree with. The inclusion of attributes provides a means of giving context to the nodes of the derivation tree, with choices of terminal or non-terminal nodes at one point in the tree being able to influence choices of nodes at others.

An AG uses two distinct types of attributes, *inherited* and *synthesised*. The names are used to indicate the direction the attributes passes information in the derivation tree. Inherited being used to identify attributes which pass information down the tree and synthesised for attribute which pass information up or across tree nodes. Semantic functions are used to interpret attribute information, using it to make decisions at one point in the tree based on values of attributes set in another. Semantic functions may also include "helper" type functions that perform more subtle analysis of attributes and help semantic function make decision on values to assign to attributes.

One of the most powerful features of GE comes from its decoupling of an individuals underlying representation from that of the derivation tree it produces. All grammar information need merely be outlined in a BNF file and GE can begin generating derivation trees. We strongly feel that any extension to GE to support attribute grammars needs to strive to maintain this decoupling and with this in mind we propose a new GE system which supports AG in addition to CFG BNF specifications.

The main core of our system was designed using C++, with the attribute information, needed to be added to derivation tree nodes being stored in shared memory space using C++ pointers. Our system includes an embedded Python interpreter used to run the semantic function and a C++/Python interface which allows semantic functions interact with attributes in shared memory.

Storing attributes in shared memory allows them to be assigned to any number of nodes in the derivation tree and also facilitates the passing of information in any direction between the nodes. Changes to an attribute at one node are immediately seen at all other nodes that share the same attribute. As Python is dynamically typed it reduces the complexity of the semantic function specifications and allows the loading of semantics at runtime rather than having to compile them separately before running the GE system. This was a carefully chosen design to help maintain in as much as possible the containment of the entire AG specification on the single BNF file, like that of a CFG.

The rest of this paper is organised as follows: Sect. 2 discusses some of the existing attribute grammar capable GE systems, discusses their use of AG and highlighting the difference of our proposed new system; Sect. 3 outlines a set of experiments carried out using our new system, using first a CFG and then an annotated version of the CFG (an AG); finally Sect. 4 concludes the paper, highlighting again the main motivations of our new GE system design and suggesting some of the areas we can extend our system into in the future.

2   Background

We are not the first to present results of experiments carried out using a GE system with added support for attribute grammars. As far back as 2005 de la Cruz et al. [1] presented results of experiments carried out on symbolic regression type problems using GE with CFG and AG specifications. More recently Karim and Ryan carried out a number of experiments using GE with AG on a variety of problem types including, but not limited to, their work on the artificial ant trail problem [3].

The results presented by both clearly demonstrate the performance gains a GE system can achieve by supporting AG problem specification. This is something which will become more important when dealing with problems with increasingly large train and test sets and ever more time consuming fitness evaluation cycles.

Both de la Cruz and Karim provided very little by way of description of their underlying GE systems design, choosing instead to only focus on the performance gains fitness gains seen in the solutions produced. Neither discusses attribute storage strategies or their effect on the information passing between the nodes of the derivation tree, or the means of specification of semantic functions and their interaction with the attributes in the derivation tree. Our system utilizes a number of features in an effort to keep the newly added AG specification as concise and clear as possible, something we feel merits highlighting a paper outlining an extension to GE to support AG.

The semantics outlined in an AG, as used by a GE system, act as a form of logic which, along with the grammar production rules, guides the generation of the derivation tree during the mapping process. Attributes can be used to pass information between tree nodes giving them a context, something that is not possible with a CFG. From a design point of view when adding support for

AG it makes sense to abstract out the logic (semantics) from the underlying representation (derivation tree) in the same way that a CFG does with the production rule specification in a BNF file. This is something we have done in our system. This will make the expression of semantics less troublesome, allowing them to be included in, and read directly from, a BNF at the same time as the production rules.

3   Experiments

We chose four symbolic regression (SR) type problems on which to test our new system. Problems 1 and 2 have a single independent input, X, while problems 3 and 4 have an additional independent input Y. Details of the problem equations, along with the range of data points used for train and test are outlined in Table 1.

Table 1. Problem sets and train and test data point ranges

	Problem	Training set [$min : step$] 50 points
		Test set [$min : step$] 200 points
1	$arcsinh(x)$	$[0.0 : 1.0]$
		$[0.1 : 0.25]$
2	$x^3 e^{-x} cos(x) sin(x)(sin^2(x)cos(x) - 1)$	$[0.0 : 0.2]$
		$[0.05 : 0.05]$
3	$y^3 e^{-x} cos(y) sin(x)(sin^2(y)cos(x) - 1)$	$x[0.0 : 0.2], y = x + 0.03$
		$x[0.05 : 0.05], y = x + 0.03$
4	$y^2 x^6 - 2.13 y^4 x^4 + y^6 x^2$	$x[1.9 : 0.075], y = x + 0.015$
		$x[1.91 : 0.019], y = x + 0.015$

3.1   Setup

An initial CFG specification was created which includes a set of basic mathematical operators $(+, -, *, /)$ and a set of 50 persistent random constants [2], PRC, generated in the range $PRC = \{c | c \in \Re \wedge -5 \leq c < 5\}$. The CFG was designed so there is a 50/50 chance of either an independent variable (X or Y) or a PRC getting added to the derivation tree. When a choice is made to add a PRC, i.e. <prc> ::= PRC, the codon value and mod operation are used to select which of the 50 available prc values to choose. For the sake of conciseness we use PRC in the grammar specification in Table 2, in the grammar used by our system this is replaced with the 50 prc values.

For our AG a set of attributes and semantic function were designed with two main goals:

1. Provide a globally accessible shared memory space, called "globalCache", which all nodes in the derivation tree have access to. When a node gets expanded to a terminal its value is added to the globalCache so any node can access the current evaluation state of the derivation tree as sub-trees become fully formed (i.e. expanded to terminals)
2. To track when an <op> Symbol node get expanded to, '/', and pass the information back up the tree so the semantic function can use it to ensure that a / is not followed by value that could be zero.

Along with providing access to add, read and update attribute information on the nodes of a derivation tree our systems C++ / Python interface also provides a means for the semantic function to directly access the Symbol information stored at node. As can be seen in the grammar specification in Table 2, using a masked property, ".Data", the semantic function can change the terminal value from "X" to "X + PRC", or "Y" to "Y + PRC" or whatever other value desired.

When each production rule is being read initially from a BNF our system tests to see if the production includes a set of additional terms, enclosed in a set of curly brackets. If the set of curly brackets is found then its contents are formatted into a Python function which is made available to the Python interpreter so it can be called during the creation of the derivation tree. We can also very easily include any other Python library, available on the system, in a semantic function, or create our own semantic helper functions designed specifically for use with a particular set of problems. In our AG we have defined a simple helper function which is included in all semantics functions. It was designed to perform a particular simple function, which is defined as follows:

appendSymbol(nodeOne.A, B). If the derivation tree node, nodeOne has an attribute called 'A', then its shared memory space is accessed and its contents is updated, appending the value 'B' to whatever already exists in it. If nodeOne does not have an attribute called 'A' then nothing is done.

50 runs were carried out for each problem, using each type of grammar, and the results presented are averaged over those runs. Normalised linear scaled mean squared error (NLSMSE) [4] was used as a fitness measure in both the sets of experiments. Details of the GE system parameters used for each run are outlined in Table 3.

We had initially hoped to include the use of the Python Abstract Syntax Tree (AST) library in our semantic functions but it was unfortunately not fully operation in this version of our system. Using the AST library we could potentially evaluate expressions as they appear in sub-tree segments of the derivation tree. This could be a very powerful feature and among other things be used to help prevent the generation of more difficult to detect invalids. It is something we hope to have implemented in the next revision of our system.

Table 2. CFG and AG specifications

			Semantics (AG only)
S	::=	\<expr\>	\<expr\>.globalCache = ' ';
\<expr$_1$\> ::=	\<expr$_2$\> \<op\> \<expr$_3$\>		\<expr$_2$\>.globalCache ← \<expr$_1$\>.globalCache
			\<op\>.globalCache ← \<expr$_1$\>.globalCache
			\<expr$_3$\>.globalCache ← \<expr$_1$\>.globalCache
			\<op\>.op = ' '
			\<expr$_3$\>.lastOp ← \<op\>.op
	| (\<expr$_2$\> \<op\> \<expr$_3$\>)		\<expr$_2$\>.globalCache ← \<expr$_1$\>.globalCache
			\<op\>.globalCache ← \<expr$_1$\>.globalCache
			\<expr$_3$\>.globalCache ← \<expr$_1$\>.globalCache
			\<op\>.op = ' '
			\<expr$_3$\>.lastOp ← \<op\>.op
	|	\<var\>	\<var\>.globalCache ← \<expr$_1$\>.globalCache
			\<var\>.lastOp ← \<expr$_1$\>.lastOp
\<op\> ::=	+		\<op\>.op = '+'
			appendSymbol(\<op\>.globalCache, '+')
	|	−	\<op\>.op = '-'
			appendSymbol(\<op\>.globalCache, '-')
	|	*	\<op\>.op = '*'
			appendSymbol(\<op\>.globalCache, '*')
	|	/	\<op\>.op = '/'
			appendSymbol(\<op\>.globalCache, '/')
\<var\> ::=	\<ind\>		\<ind\>.globalCache ← \<var\>.globalCache
			\<ind\>.lastOp ← \<var\>.lastOp
	|	\<prc\>	\<prc\>.globalCache ← \<var\>.globalCache
			\<prc\>.lastOp ← \<var\>.lastOp
\<ind\> ::=	X		if(\<ind\>.lastOp == '/'):
			....prc = getPRC()
			....X.Data = 'X + ' + prc
			....appendSymbol(\<ind\>.globalCache, X.Data)
			else:
			....appendSymbol(\<ind\>.globalCache, 'X')
	|	Y	if(\<ind\>.lastOp == '/'):
			....prc = getPRC()
			....Y.Data = 'Y +' + prc
			....appendSymbol(\<ind\>.globalCache, Y.Data)
			else:
			....appendSymbol(\<ind\>.globalCache, 'Y')
\<prc\> ::=	PRC		appendSymbol(\<prc\>.globalCache, PRC)

3.2 Results

For a given problem the same set of training and testing data points were used for both the CFG and AG runs. Table 4 outlines the experimental results which include the mean and best fitness score achieved on both the train and test data sets for each problem (± standard deviation included with each mean).

As can be seen for the results, for each problem type the run using the AG achieved better fitness scores on both the train and test data sets. While the semantics included in our AG are relatively simple they do prevent the creation of certain invalids and this can be seen to influence the resulting scores. A table

Table 3. Run configuration parameters

Population size	500
Run terminates at	150 generations
Operator probabilities	Crossover: 0.9, mutation: 0.1
Tournament size	2
Replacement	Steady state, inverse tournament
PRC	$PRC = \{c|c \in \Re \wedge -5 \leq c < 5\}$
	$|PRC| = 50$
Normalised fitness	$\frac{1}{1+LSMSE}$
Initialisation	Ramped half and half
	(max. initial depth = 8)
Max wraps	5

Table 4. Results

	Problem	1	2	3	4
CFG	Mean train	0.8784 ± 0.0736	0.9126 ± 0.0078	0.9154 ± 0.0067	0.8174 ± 0.0077
	Mean test	0.9926 ± 0.0649	0.9089 ± 0.0068	0.9103 ± 0.0069	0.8973 ± 0.0080
	Best train	0.8866	0.9205	0.9319	0.9541
	Best test	0.9940	0.9111	0.9318	0.9418
AG	Mean train	0.8895 ± 0.0728	0.9139 ± 0.0048	0.9184 ± 0.0076	0.8596 ± 0.0078
	Mean test	0.9927 ± 0.0588	0.9098 ± 0.0009	0.9131 ± 0.0087	0.8410 ± 0.0079
	Best train	0.9012	0.9282	0.9534	0.9713
	Best test	0.9943	0.9282	0.9534	0.9701

of Vargha-Delaney A measure values comparing AG to CFG for both train and test on each problem type is provided in Table 5.

Table 5. Vargha-Delaney a measure results

Problem	1	2	3	4
Train	0.5226	0.5348	0.5954	0.5456
Test	0.5108	0.5732	0.541	0.576

3.3 Discussion

While the results presented do show an improvement in overall fitness by using the AG there is room for further improvement. The semantics we used are not very sophisticated and fail to take full advantage of the storage of attributes in shared memory and semantic functions access to a number of useful Python libraries.

There are a number of areas where we feel we can improve our systems performance further, evaluating sub-tree expressions in derivation and using the information to prevent the formation of more complex and difficult to detect invalids in population, dynamic pre-processing of train inputs at run time and automatic generation of semantics based on the grammar symbols and inputs, to name a few.

Each of the problems presented had a relatively small cost associated with fitness evaluation so the effects of evaluating invalids is not very pronounced. We can however see that in situations where this is not the case the increased precision of specification provided by AG will become even more important.

4   Conclusions

In this paper we introduced a new GE system designed to support the specification of problem semantics in the form of attribute grammars. We provided a description of the underlying motivations for our system design, with a core built using C++, storage of attribute information in shared pointers and support for semantic function specification in Python scripts.

We followed this by briefly discussing some of the existing GE AG systems comparing, in as much as possible, the main goals of our system design to them and emphasising why we feel our systems design could help make the specification and use of AG with GE much more straight forward and concise.

We then outlined a set of experiments carried out using the new GE system, one using a traditional CFG and another using a relatively simple AG. We discussed the results, highlighting the performance improvements seen by using an AG and finally we finish by suggesting some of the areas we feel we can extend our system in future.

References

1. de la Cruz Echeandía, M., de la Puente, A.O., Alfonseca, M.: Attribute grammar evolution. In: Mira, J., Álvarez, J.R. (eds.) IWINAC 2005. LNCS, vol. 3562, pp. 182–191. Springer, Heidelberg (2005)
2. Dempsey, I., O'Neill, M., Brabazon, A.: Constant creation in grammatical evolution. Int. J. Innov. Comput. Appl. 1(1), 23–38 (2007)
3. Karim, M.R., Ryan, C.: Sensitive ants are sensible ants. In: Proceedings of the Fourteenth International Conference on Genetic and Evolutionary Computation Conference, pp. 775–782. ACM (2012)
4. Keijzer, M.: Improving symbolic regression with interval arithmetic and linear scaling. In: Ryan, C., Soule, T., Keijzer, M., Tsang, E., Poli, R., Costa, E. (eds.) EuroGP 2003. LNCS, vol. 2610, pp. 70–82. Springer, Heidelberg (2003)
5. Knuth, D.E.: Semantics of context-free languages. Math. Syst. Theory 2(2), 127–145 (1968)
6. O'Neill, M., Ryan, C.: Grammatical evolution: Evolutionary Automatic Programming in an Arbitrary Language. Kluwer, Boston (2003)

Indirectly Encoded Fitness Predictors Coevolved with Cartesian Programs

Michaela Sikulova[(✉)], Jiri Hulva, and Lukas Sekanina

Faculty of Information Technology, IT4Innovations Centre of Excellence,
Brno University of Technology, Božetěchova 2, 612 66 Brno, Czech Republic
{isikulova,sekanina}@fit.vutbr.cz, xhulva00@stud.fit.vutbr.cz

Abstract. We investigate coevolutionary Cartesian genetic programming that coevolves fitness predictors in order to diminish the number of target objective vector (TOV) evaluations, needed to obtain a satisfactory solution, to reduce the computational cost of evolution. This paper introduces the use of coevolution of fitness predictors in CGP with a new type of indirectly encoded predictors. Indirectly encoded predictors are operated using the CGP and provide a variable number of TOVs used for solution evaluation during the coevolution. It is shown in 5 symbolic regression problems that the proposed predictors are able to adapt the size of TOVs array in response to a particular training data set.

Keywords: Coevolution · Cartesian genetic programming · Fitness prediction

1 Introduction

The development of *Genetic Programming* (GP) is mainly driven by the increasing demand to solve complex problems which cannot be solved directly or systematically using informed methods. In many real-world applications, the fitness evaluation of a candidate program is computationally very expensive. Often, the fitness in GP is calculated over a set of *fitness cases* [11]. A fitness case corresponds to a representative situation in which the ability of a program to solve a problem can be evaluated. A fitness case consists of potential program inputs and target values expected from a perfect solution as a response to these program inputs. Potential program inputs and the corresponding target values are ordered in a sequence called *target objective vector* (TOV).

A set of TOVs (*training data*) is typically a small sample of the entire domain space. The choice of how many TOVs (and which ones) to use is often a crucial decision since whether or not an evolved solution will generalize over the entire domain depends on this choice. It also holds for the evolutionary design which has been performed by *Cartesian Genetic Programming* (CGP). In the case of symbolic regression or the evolutionary image filter design (which is one of the typical application domains for CGP [8]), from hundreds to tens of thousands TOVs have to be evaluated in order to obtain a single fitness value. In order to

© Springer International Publishing Switzerland 2015
P. Machado et al. (Eds.): EuroGP 2015, LNCS 9025, pp. 113–125, 2015.
DOI: 10.1007/978-3-319-16501-1_10

find a robust and acceptable solution a large number of fitness evaluations has to be performed.

Fitness modeling methods have been used to reduce the computational complexity of expensive fitness evaluations [2]. A predefined model or coarse-grained simulation has been used to approximate the fitness value in cases in which obtaining the exact fitness requires an expensive simulation or a physical experiment. Machine learning methods or a subsampling of training data can be used in order to approximate the fitness efficiently. However, it is not always clear when the benefits of fitness modeling can outweigh the cost.

A closely related concept to fitness modeling is *fitness prediction*, which is a technique used to replace fitness evaluations by a lightweight approximation that adapts with the solution evolution. Fitness predictors cannot approximate the entire fitness landscape, but they are instead shifting their focus throughout the evolution. An algorithm that coevolves fitness predictors, optimized for the solution population, has been introduced for standard (tree-based) genetic programming in order to reduce the fitness evaluation cost and frequency by Schmidt and Lipson [7].

In our previous work, inspired by *coevolution of fitness predictors* [7] and the *coevolutionary principles* which have been summarized in [4], we applied a coevolution of TOVs in order to accelerate fitness evaluations in CGP. We adopted the fitness predictor encoding in the form of a subset of training data. Fitness predictors have been represented as a constant-size array of pointers to elements in the training data and operated using a simple genetic algorithm. Coevolutionary algorithm has been adapted for CGP. We have obtained a significant speedup (2.03–5.45) over the standard CGP for 5 symbolic regression problems [10] and the results have been very competitive with tree-based GP. The same coevolutionary CGP and Hillis' *competitive coevolution* approach [1] adapted for CGP have been used in the evolutionary image filter design [9]. Although the median time of evolution has been reduced 2.99 times in comparison with standard CGP, a large number of experiments had to be accomplished in order to find the most advantageous size of the fitness predictor (the number of TOVs in predictor) for this particular task. An open problem is how to reduce this overhead.

This paper deals with a new type of fitness predictors whose size is changing dynamically during the coevolution. These fitness predictors with a variable number of TOVs are represented in the form of functional expressions. This functional expression generates a certain number of indexes into the training data. Indexes then address specific TOVs from the original training data which are selected for solution fitness prediction. The proposed method is evaluated using 5 symbolic regression problems and compared with the original approach.

The paper is organized as follows. Section 2 introduces Cartesian genetic programming, Sect. 3 summarizes our previous work on Coevolution of Fitness predictors in the CGP and outlines an open issue. In Sect. 4, a new approach to fitness predictor encoding is presented. Experimental results are discussed in Sect. 5. Finally, conclusions are given in Sect. 6.

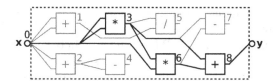

Fig. 1. A candidate program in CGP, where $l = 4, n_c = 4, n_r = 2, n_i = 1, n_o = 1, n_a = 2, \Gamma = \{+ (1), - (2), * (3), / (4)\}$ and chromosome is: 0, 0, 1; 0, 0, 1; 0, 0, 3; 2, 2, 2; 3, 1, 4; 3, 0, 3; 3, 6, 2; 3, 6, 1; 8.

2 Cartesian Genetic Programming

The state of the art of Cartesian genetic programming has been summarized in a monograph [3]. CGP is a variant of genetic programming that uses a specific encoding in the form of directed acyclic graph and a mutation-based search. CGP has been successfully employed in many traditional application domains of genetic programming such as symbolic regression, but has been predominantly applied in evolutionary design and optimization of logic networks.

A candidate program in CGP is modelled as a Cartesian grid of $n_c \times n_r$ (columns $\times$ rows) programmable elements (nodes). The number of primary inputs, n_i, and outputs, n_o, of the program are defined for a particular task. Each node input can be connected to the output of a node placed in previous l columns or to one of the program primary inputs. The types of n_a-input node functions are decided by user and defined in the set Γ. Each node of the directed graph represents a particular function and is encoded by $n_a + 1$ genes. One gene is the code of node function, the remaining genes are the indexes of the node input connections. Figure 1 shows an example of a candidate program and its encoding in the chromosome.

In CGP, a variant of a simple $(1 + \lambda)$ evolutionary algorithm is used as a search mechanism. The initial population is constructed either randomly, by a heuristic procedure or uses an existing solution. Every new population consists of the best individual of the previous generation (so-called parent) and its λ offspring. To create the offspring individuals from the parent, a point mutation operator is used. Mutation modifies h randomly selected genes to another randomly generated (but valid) values.

3 Fitness Prediction in CGP

In our previous work, fitness predictors were small subsets of the training data and coevolved with CGP programs [10]. An optimal fitness predictor was sought using a simple genetic algorithm (GA) which operated a population of fitness predictors. Fitness predictor was directly encoded as a constant-size array of pointers to the elements (TOVs) in the training data. It was shown in 5 symbolic regression benchmarks that only 12 TOVs for fitness prediction were needed to find a satisfactory solution. Moreover, a significant improvement (in terms of computational cost reduction) has been obtained in comparison with CGP without coevolution.

The coevolution adapted for CGP has been used in the evolutionary design of image filters, where the standard CGP has been successful so far. Using coevolutionary CGP, a computational cost reduction has been obtained too [9]. However, this utilization brings some potential problems. The process of finding the most advantageous setting in terms of the fitness predictor size for this particular task was the most time consuming part of the experiments. Too many independent runs had to be performed to observe that 15–20% (about 10 thousand of TOVs) of original training data are needed to find an image filter of the same quality of filtering as the best filter evolved using the standard CGP utilizing the original training data. While using GA chromosomes as long as thousands genes, the so-called *scalability problem* has been observed. In the context of EAs the scalability problem refers to the situation in which the evolutionary algorithm is able to provide a very good solution to a small problem instance, but only unsatisfactory solutions can be generated for larger problem instances.

4   Proposed Method

The number of TOVs required to obtain a satisfactory solution is variable from benchmark to benchmark. To simply apply a coevolutionary CGP to a new, unknown task, we should consider a fitness predictor with the dynamic size which can be adapted during the coevolutionary process. Although the direct encoding of the predictor involves a simpler encoding which is suitable for basic applications, more complex tasks need sizable predictors that are sorely handled by GA. Several possible encodings of fitness predictor have been mentioned in [6]. Fitness predictors, in this work are not, however, encoded as the constant-size array of TOVs. Instead, we use an indirect encoding in the form of functional expression selecting particular TOVs. TOVs used for fitness prediction are selected by means of indexes that are generated using this expression.

The evolution of this expression can be seen as a form of a symbolic regression task which is a typical task for genetic programming. We have considered to employ CGP due to a simpler and faster operation on chromosomes.

4.1   Indirectly Encoded Predictor

In this paper, the evolution of fitness predictors is based on the principles of CGP as introduced in Sect. 2. The predictor chromosome encodes a Cartesian grid of two-input functional nodes operating over one primary input and returning two primary outputs. In addition to the Cartesian grid, an initializing value x_0 is encoded in the chromosome. It is operated using a special mutation operator – value x_0 is multiplied by a randomly generated real number in the user-defined range.

While composing the array of TOVs for fitness prediction, the initializing value x_0 is used as a primary input of the predictor. In response to the primary input x_i, the candidate predictor returns two outputs - $out_0(x_i)$ and $out_1(x_i)$. Index $j(x_i)$ of selected TOV is then calculated as:

$$j(x_i) = out_0(x_i) \mod n, \tag{1}$$

where n is the total number of TOVs in the training data. TOVs selection continues with the next iteration using $out_0(x_i)$ as a new primary input of the predictor, until the $out_1(x_i)$ is out of the user-defined range r_{out1} or the maximum size of the array of TOVs for fitness prediction is reached.

4.2 Predictor Training

Predictors have to be coevolved with the solution evolution in order to adapt them to the solved problem. Predictor training data consists of *fitness trainers*, which are selected copies of candidate solutions occurred during the solution evolution and their corresponding exactly measured fitness values f_{exact} (i.e. fitness evaluated using the original training data).

The archive of trainers has a well defined structure. The number of trainers in the archive is kept constant during the evolution. While initializing the coevolution, solutions from the first generation are chosen and copied to the archive of trainers. If the archive of trainers is larger than the solution population, missing trainers are generated randomly. Trainers in the archive are updated periodically – the top-ranked candidate solution is copied to the trainers archive if its predicted fitness value differs from the top-ranked candidate solution in the previous generation; the next trainer is updated using a random solution. A new trainer t replaces the oldest one in the trainers archive and the exact fitness of the new trainer $f_{exact}(t)$ is evaluated. This approach to the predictor training data structure leads to maintaining a representative sample of the current solution population (due to the copies of top-ranked candidate solutions) as well as maximizing fitness diversity of solutions in the archive of trainers (due to the randomly generated trainers).

4.3 Fitness of Predictor

While designing the fitness function of the indirectly encoded predictor, two main options should be considered: (1) prediction precision and (2) prediction cost. Prediction precision of predictor p is calculated using the *relative error* of the exact and predicted fitness values of solutions in the trainers archive:

$$\text{prec}(p) = \frac{1}{u} \sum_{j=1}^{u} u \frac{|f_{exact}(t_j) - f_{predicted}(t_j)|}{f_{exact}(t_j) + c}, \tag{2}$$

where u is the number of solutions in the trainers archive, parameter c allows to moderate a sharp increase of relative error while the $f_{exact}(t_j)$ is very close to 0.

Prediction cost of a predictor is depended on how many TOVs have to be used while evaluating the predicted fitness. The number of TOVs in the array for

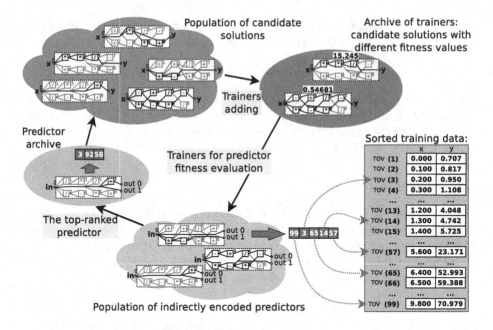

Fig. 2. Coevolution of candidate solutions and fitness predictors.

fitness prediction – size (p) – is employed for this purpose. To simplify the fitness evaluation process, both objectives are embedded in the single fitness function. To establish the fitness function for predictor p, a desired function flow has been processed by the Eureqa software [5], and the following function has been obtained:

$$f(p) = (a \cdot \text{prec}(p))^4 + b \cdot \text{size}(p) \left(1 + a \cdot \text{prec}(p)^2\right), \tag{3}$$

where size (p) is the number of TOVs addressed by predictor p. Parameters a and b control the preference between the prediction precision and the prediction speed (i.e. the number of TOVs addressed by predictor).

4.4 Coevolution of Solutions and Predictors

In the first step, the candidate solutions and candidate predictors are randomly initialized. Then the solution evolution waits for the first top-ranked predictor (obtained in the first generation of predictors). The evolution of candidate solutions is based on the principles of CGP. TOVs addressed by the predictor loaded from the *predictor archive* are used for fitness prediction of the solutions. The top-ranked predictor for solution fitness prediction is then updated periodically in a user-defined number of solution generations. While evolving solution, top-ranked solutions with different fitness are copied to the one half of the trainers archive. In each generation of predictors, one trainer from the second half of trainers archive is updated randomly.

Predictors are also evolved using CGP. Using each predictor in the current generation, the predicted fitness of trainers is evaluated and the fitness of the predictor is established. The top-ranked predictor is then updated in a predictor archive and also used for producing a new generation of predictors. The overall scheme of the proposed coevolutionary algorithm is shown in Fig. 2.

If a satisfactory solution is found or the user-defined maximum number of solution generations is reached, the coevolution is terminated.

5   Results

This section presents benchmark problems, experimental setup and experimental evaluation of the proposed approach and its comparison with the original directly encoded fitness predictors and standard CGP without coevolution.

5.1   Benchmark Problems

Five symbolic regression benchmark functions (F1–F5) were selected as TOV sources for evaluation of the proposed method:

$$F1 : f(x) = x^2 - x^3, \qquad\qquad x = [-10 : 0.1 : 10]$$
$$F2 : f(x) = e^{|x|} \sin(x), \qquad\qquad x = [-10 : 0.1 : 10]$$
$$F3 : f(x) = x^2 e^{\sin(x)} + x + \sin\left(\frac{\pi}{x^3}\right), \qquad x = [-10 : 0.1 : 10]$$
$$F4 : f(x) = e^{-x} x^3 \sin(x) \cos(x) \left(\sin^2(x) \cos(x) - 1\right), \quad x = [0 : 0.05 : 10]$$
$$F5 : f(x) = \frac{10}{(x-3)^2 + 5}, \qquad\qquad x = [-2 : 0.05 : 8].$$

In order to form a training data, 200 equidistant distributed samples were taken from each function. Functions F1, F2 and F3 are taken from [7], functions F4 and F5 from [12] and all functions F1–F5 were used in order to evaluate coevolution of CGP and directly encoded predictors [9].

5.2   Experimental Setup

The setup of the solution evolution is used according to literature [10], i.e. $\lambda = 12, n_i = 1, n_o = 1, n_c = 32, n_r = 1, l = 32$, every node has two inputs $(i_1, i_2), \Gamma = \{i_1 + i_2, i_1 - i_2, i_1 \cdot i_2, \frac{i_1}{i_2}, \sin(i_1), \cos(i_1), e^{i_1}, \log(i_1)\}$ and the maximum number of mutation per individual is $h = 8$. The solution fitness function is defined as the relative number of hits. There are, in fact, two fitness functions for candidate solution s. While the exact fitness function $f_{exact}(s)$ utilizes the complete training set, the fitness function for fitness prediction $f_{predicted}(s)$ employs only selected TOVs. Formally,

$$f_{exact}(s) = \frac{1}{n} \sum_{j=1}^{n} g(y(j)) \tag{4}$$

$$f_{predicted}(s) = \frac{1}{m} \sum_{j=1}^{m} g(y(j)) \tag{5}$$

$$g(y(j)) = \begin{cases} 0 \text{ if } |y(j) - t(j)| \geq \varepsilon \\ 1 \text{ if } |y(j) - t(j)| < \varepsilon \end{cases} \tag{6}$$

where $y(j)$ is a candidate program response to TOV j, $t(j)$ is the target response, n is the number of TOVs in the training data, m is the number of TOVs addressed by predictor and ε is a user-defined acceptable error – for benchmarks F1, F2: 0.5; F3: 1.5; F4, F5: 0.025. The acceptable number of hits is 97 %.

To find the most advantageous setting of the predictor evolution, over 160,000 independent runs were performed. The results were obtained using the following setup of the predictor evolution: $\lambda = 4, n_i = 1, n_o = 2, n_c = 15, n_r = 2, l = 4$, every node has two inputs (i_1, i_2), $\Gamma = \{i_1+i_2, i_1-i_2, i_1 \cdot i_2, \frac{i_1}{i_2}, \sin(i_1), \max(i_1, i_2),$ $\min(i_1, i_2), -i_1, i_1 \mod i_2, |i_1|$ and number of mutation per individual is $h = 30$. The range of out_1 (affecting the number of TOVs addressed by predictor) is set as $-1000 < r_{out1} < 1000$; the minimum number of TOVs addressed by predictor is 5 $(2, 5\%$ of the complete training data) and the maximum number is 50 (25%). Parameters of predictor fitness function were empirically set as follows: Predictor precision (Formula 2) parameter $c = 0.002$ and the predictor fitness function (Formula 3) parameters $a = 17$ and $b = 0.04$. Every 2,000 generations of the solution evolution, a new predictor has been loaded for the solution fitness evaluation.

5.3   Comparisons of the Algorithms

The goal of this experiment is to compare the proposed coevolution of indirectly encoded fitness predictors evolved using CGP (FP_{CGP}) with the original directly encoded fitness predictors evolved using GA (FP_{GA}) and standard CGP without coevolution (CGP_{STD}). In all the algorithms, solutions are evolved using the equivalent setup as presented in Sect. 5.2. FP_{GA} is used according to literature [9], i.e. 12 TOVs in chromosome, 32 individuals in predictor population, 2-tournament selection, a single point crossover and the mutation probability 0.2. The algorithms are compared in terms of the success rate (the number of runs, giving a solution with predefined quality), the number of generations and the number of TOV evaluations to converge (in order to compare the computational cost). Table 1 gives the median values calculated of 50 independent runs for each benchmark function F1–F5.

It can be seen from Table 1 that both coevolutionary approaches have reached a satisfactory solution using a significantly fewer TOV evaluations than the standard CGP. Despite the fact that during FP_{CGP} evolution predictors with a large number of TOVs have to be evaluated, the number of TOV evaluations

Table 1. Comparison of standard CGP (CGP_{STD}) and coevolutionary CGP with directly encoded predictors FP_{GA} and indirectly encoded predictors FP_{CGP}.

	Algorithm	F1	F2	F3	F4	F5
Success rate	CGP_{STD}	100 %	100 %	100 %	80 %	24 %
	FP_{GA}	100 %	100 %	100 %	100 %	100 %
	FP_{CGP}	100 %	100 %	100 %	100 %	90 %
Generations to converge (median)	CGP_{STD}	$1.11 \cdot 10^3$	$4.46 \cdot 10^3$	$1.76 \cdot 10^5$	$7.15 \cdot 10^5$	$1.36 \cdot 10^6$
	FP_{GA}	$2.62 \cdot 10^3$	$2.53 \cdot 10^3$	$1.10 \cdot 10^5$	$1.00 \cdot 10^6$	$1.34 \cdot 10^6$
	FP_{CGP}	$1.00 \cdot 10^3$	$2.25 \cdot 10^3$	$4.11 \cdot 10^4$	$1.47 \cdot 10^6$	$1.74 \cdot 10^6$
TOV evaluations to converge (median)	CGP_{STD}	$2.68 \cdot 10^6$	$1.08 \cdot 10^7$	$4.24 \cdot 10^8$	$1.72 \cdot 10^9$	$3.28 \cdot 10^9$
	FP_{GA}	$5.20 \cdot 10^5$	$5.01 \cdot 10^5$	$2.19 \cdot 10^7$	$2.00 \cdot 10^8$	$2.67 \cdot 10^8$
	FP_{CGP}	$7.43 \cdot 10^5$	$1.60 \cdot 10^6$	$1.90 \cdot 10^7$	$8.05 \cdot 10^8$	$8.78 \cdot 10^8$

to converge is similar for both FP_{CGP}-evolved and FP_{GA}-evolved predictors. Although CGP_{STD} evaluates the whole TOVs set in every fitness function call, the number of generations is comparable for all three methods.

5.4 Predictor Behaviour

In this section we discuss how the predictors are able to select a representative sample of TOVs which allows for obtaining a satisfactory solution. However, it should be pointed out that to facilitate an indirectly encoded predictor to maintain eventual geometries or peaks and valleys in training data, the training set should be well sorted (if it is possible).

In order to observe the behaviour of predictor data samples, we plot (see Figs. 3 and 4) the number of TOVs and the frequency of TOVs addressed by predictors, which were used during the course of evolution for solution fitness prediction (50 independent runs considered). It can be seen from Table 1 that the satisfactory solution for benchmarks F1 and F2 can be obtained by $2 \cdot 10^3$ generations of the solution fitness prediction, which is the time when only the first co-evolved predictor is ready for solution fitness prediction. Then Fig. 3 shows the number and the frequency of TOVs addressed by the top-ranked predictor taken from the very first (randomly generated) generation. While the evolution wasn't allowed to suit to the training data, sizable predictors were selected in order to accomplish a better prediction precision – see Fig. 3b and d. Despite this fact, satisfactory solutions for benchmarks F1 and F2 have been found using a comparable number of TOV evaluations (and less number of generations of solutions) in comparison with directly encoded predictors using only 12 TOVs for solution fitness prediction (FP_{GA}).

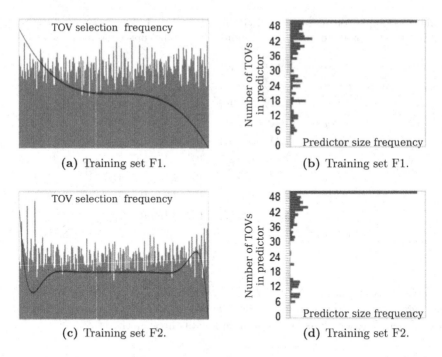

Fig. 3. Frequency and the number of TOVs in predictors used for candidate solution fitness prediction (training set F1 and F2).

In case of benchmarks F3, F4 and F5, coevolution exhausted many more generations to converge. Then predictors were able to adapt to the training data. Figure 4b, d and f shows that each benchmark prefers a different number of TOVs for fitness prediction. It can be seen from Fig. 4f that about 10 TOVs and about 38 TOVs were preferred in order to predict the fitness of candidate solutions while solving the benchmark F5.

It can be seen in Fig. 4a, c and e that sample points do not focus entirely on the peaks and valleys of the training data, but are well distributed over the data set during the coevolution, however some geometries have been observed. If all TOVs addressed by predictor focus on the interesting regions (peaks and valleys) of the training data, the predictor would represent the maximum error (which is improper while requiring the predicted fitness corresponding to the exact fitness). Furthermore, TOVs addressed by the fitness predictors are variable in response to the solution evolution. The solution evolution forces the predictors to contain two types of TOVs, some of them are easy others difficult for particular solutions.

The number of TOVs addressed by predictors also changes during the coevolution in response to the course of solution evolution. Figure 5 shows the exact fitness of the top-ranked candidate solutions during the course of coevolution and the size of the predictor used to predict their fitness during a typical run for the benchmark F3.

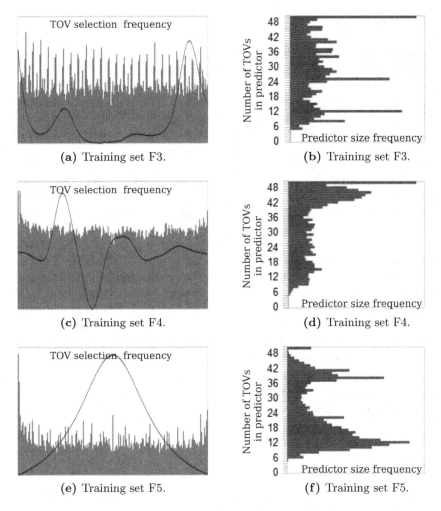

Fig. 4. Frequency and the number of TOVs in predictors used for candidate solution fitness prediction (training set F3–F5).

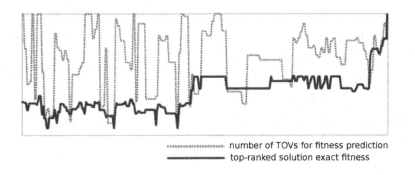

Fig. 5. Exact fitness of top-ranked candidate solutions during the course of evolution and the size of predictor during a typical run for the F3 data set.

6 Conclusions

In summary, we have introduced the use of coevolution of fitness predictors in CGP with a new type of indirectly encoded predictors. Indirectly encoded predictors are operated using the CGP and provide a variable number of TOVs used for solution fitness prediction during the coevolution in response to the solved problem. When applied to the symbolic regression problem, this approach was found to be comparable with the original directly encoded predictors using just 12 TOVs for the solution fitness prediction in terms of the number of evaluated TOVs to converge. We have shown using 5 benchmarks that proposed predictors are able to adapt the size of TOVs array for solution fitness prediction in response to the particular training data. This property enables to use the coevolution of fitness predictors for solving a new, unknown task, without the need to find the most advantageous size of the TOVs array experimentally.

However, as symbolic regression has not been considered as a typical application domain for CGP, our future work will be devoted to the utilization of the proposed fitness prediction algorithm in the evolutionary image filter design where the original directly encoded predictors have been successful so far. Considering the fact that the evolutionary design using CGP has been successfully accelerated in *field programmable gate array* (FPGA), another goal will be to implement coevolutionary CGP with indirectly encoded predictors to FPGA and thus accelerate the search process and use it in a real-world application.

Acknowledgments. This work was supported by the Czech science foundation project 14-04197S, the Brno University of Technology project FIT-S-14-2297 and the IT4Innovations Centre of Excellence CZ.1.05/1.1.00/02.0070.

References

1. Hillis, W.D.: Co-evolving parasites improve simulated evolution as an optimization procedure. Physica D **42**(1), 228–234 (1990)
2. Jin, Y.: A comprehensive survey of fitness approximation in evolutionary computation. Soft Comput. J. **9**(1), 3–12 (2005)
3. Miller, J.F.: Cartesian Genetic Programming. Springer, Heidelberg (2011)
4. Popovici, E., Bucci, A., Wiegand, R., De Jong, E.: Coevolutionary principles. In: Rozenberg, G., Bäck, T., Kok, J.N. (eds.) Handbook of Natural Computing, pp. 987–1033. Springer, Heidelberg (2012)
5. Schmidt, M., Lipson, H.: Distilling free-form natural laws from experimental data. Science **324**(5923), 81–85 (2009)
6. Schmidt, M.D., Lipson, H.: Co-evolving fitness predictors for accelerating and reducing evaluations. In: Riolo, R., Soule, T., Worzel, B. (eds.) Genetic Programming Theory and Practice IV. Genetic and Evolutionary Computation, vol. 5, pp. 113–130. Springer, Ann Arbor (2006)
7. Schmidt, M.D., Lipson, H.: Coevolution of fitness predictors. IEEE Trans. Evol. Comput. **12**(6), 736–749 (2008)

8. Sekanina, L., Harding, S.L., Banzhaf, W., Kowaliw, T.: Image processing and CGP. In: Miller, J.F. (ed.) Cartesian Genetic Programming, pp. 181–215. Springer, Heidelberg (2011)

9. Sikulova, M., Sekanina, L.: Acceleration of evolutionary image filter design using coevolution in Cartesian GP. In: Coello, C.A.C., Cutello, V., Deb, K., Forrest, S., Nicosia, G., Pavone, M. (eds.) PPSN 2012, Part I. LNCS, vol. 7491, pp. 163–172. Springer, Heidelberg (2012)

10. Šikulová, M., Sekanina, L.: Coevolution in Cartesian genetic programming. In: Moraglio, A., Silva, S., Krawiec, K., Machado, P., Cotta, C. (eds.) EuroGP 2012. LNCS, vol. 7244, pp. 182–193. Springer, Heidelberg (2012)

11. Vanneschi, L., Poli, R.: Genetic programming – introduction, applications, theory and open issues. In: Rozenberg, G., Bäck, T., Kok, J.N. (eds.) Handbook of Natural Computing, pp. 709–739. Springer, Heidelberg (2012)

12. Vladislavleva, K.: Toy benchmarks. In: Symbolic Regression: Function Discovery and More (2011). http://www.symbolicregression.com/?q=toyProblems

Tapped Delay Lines for GP Streaming Data Classification with Label Budgets

Ali Vahdat, Jillian Morgan, Andrew R. McIntyre, Malcolm I. Heywood[(✉)],
and A. Nur Zincir-Heywood

Faculty of Computer Science, Dalhousie University Halifax,
Halifax, NS B3H 4R2, Canada
ali.vahdat@dal.ca, mheywood@cs.dal.ca

Abstract. Streaming data classification requires that a model be available for classifying stream content while simultaneously detecting and reacting to changes to the underlying process generating the data. Given that only a fraction of the stream is 'visible' at any point in time (i.e. some form of window interface) then it is difficult to place any guarantee on a classifier encountering a 'well mixed' distribution of classes across the stream. Moreover, streaming data classifiers are also required to operate under a limited label budget (labelling all the data is too expensive). We take these requirements to motivate the use of an active learning strategy for decoupling genetic programming training epochs from stream throughput. The content of a data subset is controlled by a combination of Pareto archiving and stochastic sampling. In addition, a significant benefit is attributed to support for a tapped delay line (TDL) interface to the stream, but this also increases the dimensionality of the task. We demonstrate that the benefits of assuming the TDL can be maintained through the use of oversampling without recourse to additional label information. Benchmarking on 4 dataset demonstrates that the approach is particularly effective when reacting to shifts in the underlying properties of the stream. Moreover, an online formulation for class-wise detection rate is assumed, where this is able to robustly characterize classifier performance throughout the stream.

Keywords: Streaming data classification · Non-stationary · Class imbalance · Benchmarking

1 Introduction

Incremental learning from streaming data represents a new challenge for algorithms applied to classification tasks [11,12,15,17]. Such tasks are non-stationary (the underlying process creating the data changes over the course of the stream), have limited capacity for revisiting previously encountered data (single pass constraint), generally present a very imbalanced class distribution (care of the sliding window access to the data) and are subject to a labelling budget (it is prohibitively expensive to label the stream).

© Springer International Publishing Switzerland 2015
P. Machado et al. (Eds.): EuroGP 2015, LNCS 9025, pp. 126–138, 2015.
DOI: 10.1007/978-3-319-16501-1_11

In this work we revisit a general architecture previously proposed for applying genetic programming (GP) to streaming data classification tasks under label budgets [20]. The framework assumes a non-overlapping window interface (of length L) to the stream consisting of a continuous sequence of $x(t)$ to $x(t-L-1)$ exemplars. A subset of instances are stochastically sampled from the current window location and placed into a data subset (DS). Only these exemplars have their label information requested. A training cycle is then performed relative to the current DS content. The authors make use of Pareto archiving to prioritize exemplars within the DS for replacement [20]. Thus, the next time the DS content is updated (corresponding to a new window location) non-dominated exemplars can be prioritized for retention. Performance was compared to that of an Adaptive Naive Bayes (ANB) classifier that made similar assumptions regarding how to sample exemplars from the stream under a label budget [19].

In this work we undertake a through re-evaluation of the relation between DS updating and training epochs of GP. The hypothesis pursued here is that more than one generation may be performed per update to DS content without changing the label budget or provoking symptoms of over-learning. This is particularly important when attempting to support a tapped delay line (TDL) interface to the stream. Specifically, utilizing a delay line implies that when classifying exemplar $x(t)$, access to a sequence of lagged instances is supported or: $x(t - l\tau), \ldots, x(t - 2\tau), x(t - \tau), x(t)$ where l is the depth of the delay line and τ is the skip size.[1] However, each x is a vector of d attributes. Thus, the TDL implies that each exemplar is a matrix of $d \times l$ attributes. On the one hand this provides a mechanism for capturing temporal properties potentially useful for characterizing exemplar t. Conversely, the dimensionality of the input space has now undergone a significant increase, potentially making the task of learning appropriate models that much more difficult.

The final aspect developed by this work is with respect to benchmarking. In particular, results for streaming classification tasks are generally expressed in terms of the prequential error metric [4,13]. Such a metric is incrementally estimated and describes classifier performance as a trajectory through the stream. The most significant drawback of such a metric, however, is that it represents an accuracy metric. Unlike approaches to offline batch performance evaluation, it is not possible to control the distribution of exemplars throughout the (stream) data set. Hence, local regions of the stream are very likely to be imbalanced, leading to low levels of class mixing and potentially degenerate classifier behaviours. Such behaviours are not identified by accuracy style metrics typically assumed for benchmarking streaming classifiers. In this work we therefore make use of an online definition for the class-wise detection metric [15].

2   Related Work

Classification under streaming tasks has been addressed from the perspective of online and ensemble learners for a considerable period of time [11,12,15,17].

---

[1] Note that the label information is limited to that of $x(t)$ alone.

However, comparatively few works have proposed explicitly evolutionary computation (EC) frameworks for streaming data classification. Most EC effort has been placed on dynamic optimization tasks where the objective is to accurately track movement in multi-modal optimization tasks [7]. As such, label information is freely available and there is a much tighter coupling between representation space and search space. Several GP researchers have considered the case of evolving models against a completely labelled stream [2,3,5,10]. However, other than [20], only non-evolutionary ML researchers have considered the case of explicitly developing models under label budgets e.g., [16,18].

3   StreamSBB

EC frameworks proposed for dynamic optimization tasks in general make use of several key components of which support for evolvability, diversity and memory appear to be critical [7,15]. In this work we are interested in further developing a recently proposed framework for applying GP to streaming classification tasks under label budgets [20]. StreamSBB assumes a symbiotic bid-based (SBB) formulation for GP, thus solutions are coevolved into teams of programs [8]. Such modularity is necessary in non-stationary tasks to provide the basis for addressing the evolvability / plasticity requirement [15]. Moreover, multi-class classification is a natural artifact of the SBB architecture. Figure 1 summarizes the framework assumed for applying GP to non-stationary streaming tasks.

StreamSBB assumes a non-overlapping window as the generic interface to the stream $(Win(i))$. For each unique window location (i), Gap exemplars are

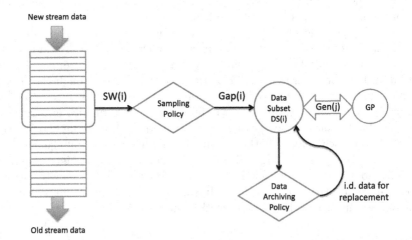

Fig. 1. StreamGP architecture. $Win(i)$ denotes the location of the non-overlapping window interface to the stream. A sampling policy selects $Gap(i)$ exemplars from $Win(i)$ without prior knowledge of label information. $DS(i)$ is the data subset (with labels) against which GP training epochs are performed. The archiving policy defines which exemplars are retained as $i \rightarrow i + 1$.

sampled with uniform probability *after which* their corresponding label is requested. Hence, a label budget is enforced. A Pareto archiving policy with an age/diversity maintenance heuristic prioritizes *Gap* exemplars for replacement from the Data Subset (DS) [1,20]. At this point ρ generations are performed where previous work assumed $\rho = 1$. As noted in the introduction, such a constraint impacts on the capacity of StreamSBB to react to changes in the stream. Note also that for $\rho > 1$ there is no change to DS content. Only when the position of the non-overlapping window to the stream shifts to the next 'chunk' of stream data is there an update to DS content. Hence, the index (i) for window location, label requests, and updates to DS content are all the same.

On the face of it, this is a relatively minor modification to the StreamSBB framework. However, as will become apparent from Sect. 5 the selection of $\rho > 1$ has a significant impact on the overall performance of the algorithm. Moreover, the significance of this is all the more pressing when classification of exemplar t is performed with reference to a tapped delay line (TDL) describing a sequence of lagged instances from the stream, i.e. a potentially more descriptive representation (see discussion from Sect. 1). The input now takes the form of a $d \times l$ matrix (# attributes by #TDL taps), thus GP needs to rationalize what specific instances to utilize while the stream continues to pass by.

Finally, anytime operation (as in predicting labels for stream content) is supported using the current $DS(i)$ content[2] to define a 'champion' individual. A class-wise detection rate metric is estimated across $DS(i)$ content for the non-dominated SBB teams alone (Sect. 4.3, Eq. (2) estimated across $DS(i)$). The latter constraint providing some robustness to selecting degenerate SBB teams. That is to say, $DS(i)$ content is a function of classes sampled from the stream under a finite label budget, thus could potentially only consist of exemplars from a single class.

4   Experimental Methodology

4.1   Streams/Datasets

Four datasets are used for the purposes of benchmarking, two of which are artificially generated and two represent real world tasks that potentially contain non-stationary properties. The reason to include two artificially generated streams is that with real-world datasets the degree of non-stationarity is unknown. Thus, artificially generated streams enable different forms of concept change to be embedded.

The artificial data streams[3] are denoted "Gradual Concept Drift" and "Sudden Concept Shift" and the real-world datasets are "Electricity Demand" [14] and "Churn Detection". Electricity Demand has frequently been employed for

---

[2] The only source of labelled data.

[3] http://web.cs.dal.ca/~mheywood/Code/SBB/Stream/StreamData.html.

Table 1. Benchmarking dataset properties. d denotes dimensionality, N refers to the total exemplar count, k reflects class count of datasets, 'Change' denotes the % of the stream in which a label change occurs.

Stream/Dataset	d	N	k	$\approx$Class Distrib. (%)	Change
Gradual Concept Drift (drift)	10	150,000	3	[16, 74, 10]	40 %
Sudden Concept Shift (shift)	6	6,500,000	5	[37, 25, 24, 9, 4]	73 %
Electricity Demand (electricity)	8	45,312	2	[58, 42]	15 %
Churn Detection (churn)	16	1,669,593	2	[91, 9]	11 %

streaming data benchmarking tasks [15] whereas the second represents an online video gaming churn prediction task. The basic properties of the datasets are summarized by Table 1 in which the last column (or 'Change') captures the frequency of label changes through the stream.

Gradual Concept Drift stream [9]: Hyperplanes are defined in a 10-dimensional space. Initial values of the hyperplane parameters are selected with uniform probability. This Dataset has 150,000 exemplars and every 1,000 exemplars, half of the parameters are considered for modification with a 20 % chance of change, hence creating the gradual drift of class concepts. Class labels are allocated as a function of hyperplanes exceeding a class threshold.

Sudden Concept Shift stream [21]: The stream is created 'block-wise' with 13 blocks and each block consists of 500,000 exemplars. Consider a concept generator tuple of the form: $\langle C1\,\%, C2\,\% \rangle$ where C1 and C2 represent two independent rule sets defining 5 class tasks. A stream is now defined in terms of the transition of exemplars from 100 % $C1$ to 100 % $C2$ in 10 % increments: $\langle 100, 0 \rangle$, $\langle 100, 0 \rangle$, $\langle 100, 0 \rangle$, $\langle 90, 10 \rangle$, $\langle 80, 20 \rangle$, ... $\langle 0, 100 \rangle$. A uniform p.d.f. is used to determine exemplar sequencing in each block.

Electricity Demand characterizes the rise and fall of electricity demand in New South Wales, Australia, using consumption and price information for the target and neighbouring regions [14]. As such it is a two class dataset (demand will either increase or decrease relative to the previous period).

Churn Detection determines the loyalty of a player toward the online video game he/she is playing. There are 16 features describing each turn of the game as well as some player-related features. The label indicates whether the player will (or not) churn within the upcoming 24–48 h time window (horizon). The set is quite unbalanced with about 91 % being class 0 (will not churn) players who will keep playing after the time window and only about 9 % being class 1 (will churn), i.e. most of the time players do not churn within the 2 day period.

4.2   Parameterization of GP

For a test stream of S_{max} exemplars a non-overlapping window of length S_{max}/i_{max} exemplars is assumed where i_{max} are the number of window locations.

The remainder of the stream passes through at a constant rate. The window content defines the pool from which the new $Gap(i)$ training exemplars are sampled and labels requested (Fig. 1).

Model initialization is performed using the first $S_{init}\%$ of the stream during the first $i_{init}\%$ of generations. Given that the interface to the stream assumed by StreamSBB is a non-overlapping window, then this defines the initial window length and implies that $i_{init}\%$ of the generations are performed against this window location. Thereafter, the sliding window advances at a fixed rate through the stream. Both S_{init} and i_{init} parameters are set to 10 percent.

Label budget is the ratio of exemplars whose labels are requested to the total stream length, or:

$$label\ budget\ (LB) = \frac{i_{max} \times |Gap|}{S_{max}} \tag{1}$$

In other words only $i_{max} \times |Gap|$ exemplars are sampled in a stream of length $S_{max}(\equiv N)$; whereas the total number of generations performed is: $i_{max} \times \rho$. Only $|Gap| = 20$ exemplars are added to $DS(i)$ (by the Sampling Policy) at each window location, $Win(i)$; hence, the label budget in the specific case of the concept shift data set would be: $LB = \frac{1,000 \times 20}{6,500,000} \approx 0.3\%$.

Given the variation in stream lengths of the benchmarking datasets (Table 1), different parameterizations for i_{max} will be assumed per dataset (summarized by Table 2). Note that i_{max} is taken to include the pre-training budget $i_{init}\%$. Table 3 summarizes the remaining generic SBB parameter settings assumed in this study e.g., population size, variation and selection operator frequencies.

Table 2. Stream dataset max. window count (i_{init}) and label budgets (LB)

Stream/Dataset	S_{max}	i_{max}	LB
Gradual Concept Drift (drift)	150,000	500	6.7%
Sudden Concept Shift (shift)	6,500,000	1,000	0.3%
Electricity Demand (elec)	45,312	500	22.1%
Churn Detection (churn)	1,669,593	1,000	1.2%

Parameterization of TDL ($x(t-l\tau), \ldots, x(t-2\tau), x(t-\tau), x(t)$) defines the length ($l$) and skip size ($\tau$) assumed or $tapSize$ and $tapSkip$ respectively. The range used for $tapSize$ is $[0, \ldots, 7]$, and $tapSkip$ is defined as $[1, 2, 4, 8, 16]$. Naturally, setting $tapSize = 0$ implies that classification is performed relative to $x(t)$ alone. Small values for $tapSkip$ imply more locality (greater resolution) whereas larger values increase the range covered by the TDL albeit at a lower resolution.

DS oversampling (or simply referred to as **oversampling**) reflects the ability of StreamSBB to decouple the rate at which GP training epochs are performed

Table 3. Generic SBB parameters. Symbiont population varies dynamically, hence no size parameter is defined. SBB assumes a 'breeder' model of evolution in which M_{gap} hosts are removed per generation [8].

Parameter	Value	Parameter	Value
Data Subset (DS) size	120	Host pop. size (M_{size})	120
Prob. symbiont deletion (p_d)	0.3	Prob. symbiont addition (p_a)	0.3
Prob. action mutation (μ_a)	0.1	Max. symbionts per host (ω)	20
Host pop. gap size (M_{gap})	60	Data Subset gap size (Gap)	20

from the rate at which the data subset content is updated (Sect. 3). The degree of oversampling is parameterized as follows: $\rho \in \{1, 2, 5\}$; where $\rho = 1$ implies one GP training epoch per data subset.

4.3 Detection Rate for Stream Data

An online class-wise **detection rate** provides the basis for incrementally estimating detection rate throughout the stream while being robust to class imbalance (unlike accuracy or error style metrics) [15]. This is particularly important under streaming data situations as models are updated incrementally and therefore sensitive to the distribution of current window content (typically a skewed distribution of classes even when the overall class distribution is balanced). The incremental class-wise detection rate can be estimated directly from stream content as follows:

$$DR(t) = \frac{1}{C} \sum_{c=[1,...,C]} DR_c(t) \text{ where } DR_c(t) = \frac{tp_c(t)}{tp_c(t) + fn_c(t)} \qquad (2)$$

where t is the exemplar index, and $tp_c(t)$, $fn_c(t)$ are the respective running totals for true positive and false negative rates up to this point in the stream.

4.4 Comparator Model

The comparator classifier is documented in a recent study of streaming data classification under label budgets and drift detection [18], and has been made available in the Massive Online Analysis (MOA) toolbox.[4] Specifically, the **Adaptive Naive Bayes (ANB)** classifier with budgeted active learning and drift detection. The 'random' active learning strategy was selected as it provided the baseline in [18] and is closest to the stochastic sampling policy adopted in this work. Active learning with budgeting is managed under a random exemplar selection policy in which stream data is queried for labels with frequency set by the budget parameter.

---

[4] MOA prerelease 2014.03; http://moa.cms.waikato.ac.nz/overview/.

5 Results

Benchmarking is performed in three phases in order to assess: (1) the contribution from tapped delay lines, (2) the role of oversampling, and (3) the performance relative to the adaptive Naive Baysian framework. All StreamSBB results represent the average of 50 runs, thus any StreamSBB performance curve is an averaged curve.

5.1 Experiments

Tap delay line (TDL) experiment: As per Sect. 4.2, between 1 and 7 historical instances can be attached to the current instance (*tapSize*). Such instances can be offset by 1 to 16 instances far from current instance (*tapSkip*).

Figure 2 illustrates how increasing tap size from 1 to 3 to 7 collectively decreases the detection rate of StreamSBB under the concept shift stream regardless of tap skip. The original StreamSBB without TDL provides an indication of the baseline performance (black dashed curve). The other datasets follow a similar trend of diminishing detection rate while increasing tap size. Varying the tap skip value provides greater history in the samples retained within the delay line. The electricity demand dataset was the only data set to respond particularly favourably to increases to skip size (page limit precludes a supporting figure).

DS oversampling experiment: Figures 3 and 4 illustrate the impact of oversampling in terms of detection rate curves for concept drift and shift streams. Higher detection rates are now maintained throughout the stream. The higher rate of oversampling appears to be preferable throughout. Further increases to the oversampling (say to a factor of 10) has only marginal positive effects (results not shown for clarity). To confirm the significance of the difference between each pair of curves, the nonparametric Mann-Whitney U test[5] of the null hypothesis is assumed (i.e. does not assume a normal distribution for data). A significance level of 0.01 is assumed and report the p-value of the test when the null hypothesis is rejected ($h = 1$) in Table 4. In all cases the higher rate of oversampling is preferred. Moreover, applying a Bonferroni Correction of $0.01/3$ to the p-values does not change this conclusion. Results for the real-world datasets were also positive and will be reported later when we compare with the ANB framework for streaming classification.

Table 4. Mann Whitney U test p-values for comparing different pairs of curves. $\times 1$, $\times 2$ and $\times 5$ define 1, 2 and 5 generations per DS update respectively.

Stream/Dataset	$\times 1$ vs. $\times 2$	$\times 1$ vs. $\times 5$	$\times 2$ vs. $\times 5$
Gradual Concept Drift	3.84e-9	5.02e-17	6.30e-13
Sudden Concept Shift	3.92e-8	1.37e-17	9.92e-13

---

[5] Also referred to as the Wilcoxon rank-sum test or Wilcoxon-Mann-Whitney test.

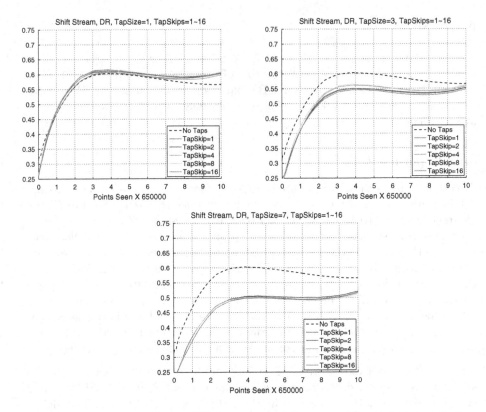

Fig. 2. Detection rate of StreamSBB for concept shift stream. The dashed curve is the case of no TDL (i.e. original StreamBB) and the solid colored curves are StreamSBB using TDL with different *tapSize* and *tapSkip* parameters.

Combined TDL and DS oversampling: Figures 5 and 6 show detection rate curves corresponding to different configurations of StreamSBB with or without TDL and/or DS oversampling. Note that for clarity the tap skip is fixed to 16 for all curves using TDL. Choosing other tap skip values returns almost identical curves for all but the electricity demand dataset.

The black solid curve is the original StreamSBB baseline before using TDL or DS oversampling. The three red curves show the diminishing trend of detection rate as tap size increases (oversampling disabled). The two blue curves show the detection rate curves when an oversampling rate of 2 and 5 are applied (TDL disabled). The black dashed curve is the detection rate curve when a tap delay line with tap size 1 and DS oversampling of rate 5 is enabled. We call this the *optimal StreamSBB* configuration. It is evident that detection rate improves once both properties are enabled. Table 5 reports the *p*-values of the Mann Whitney U test when comparing the optimal StreamSBB (using tap size 1 and oversampling rate of 5) with the original StreamSBB (disabling TDL and DS oversampling) under 0.01 significance level.

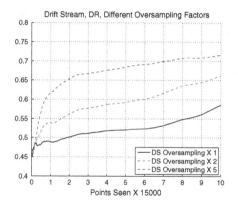

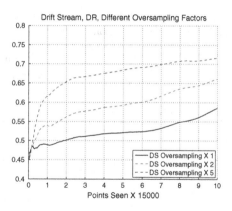

Fig. 3. DR on **drift** stream. Black: default sampling; Red: ×2 oversampling; and Blue: ×5 oversampling (Color figure online).

Fig. 4. DR on **shift** stream. Black: default sampling; Red: ×2 oversampling; and Blue: ×5 oversampling (Color figure online).

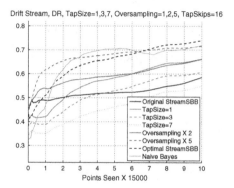

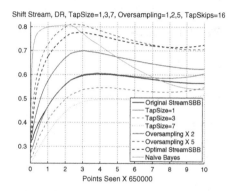

Fig. 5. DR on **drift** stream. TDL and oversampling experiments (Color figure online).

Fig. 6. DR on **shift** stream. TDL and oversampling experiments (Color figure online).

In summary, the above experiments demonstrated that a tap size of 1 outperformed a tap size 3 and 7, and the DS oversampling rate of 5 yielded statistically significantly better results than an oversampling rate of 2. A tap skip of 16 was used to take advantage of farthest historical instance for electricity dataset. Under these parameter settings the detection rate for the original StreamSBB, optimal StreamSBB and the Adaptive Naive Bayes (ANB) classifier are displayed in Figs. 7, 8, 9, and 10.

It appears that ANB has problems when there are sudden changes to the content of the stream (Fig. 8), whereas both algorithms are effective under the drift stream (Fig. 7). In all cases ANB is also able to return better detection rates much faster than StreamSBB. However, during the course of the stream

Table 5. Mann Whitney U test p-values for comparing Optimal StreamSBB and Original StreamSBB detection distributions.

Data set	Drift	Shift	Electicity	Churn
p-value	8.46e-18	1.08e-17	7.07e-18	7.06e-18

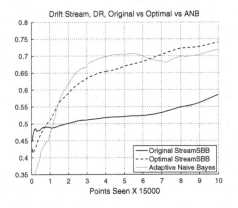

Fig. 7. DR curves for Concept Drift

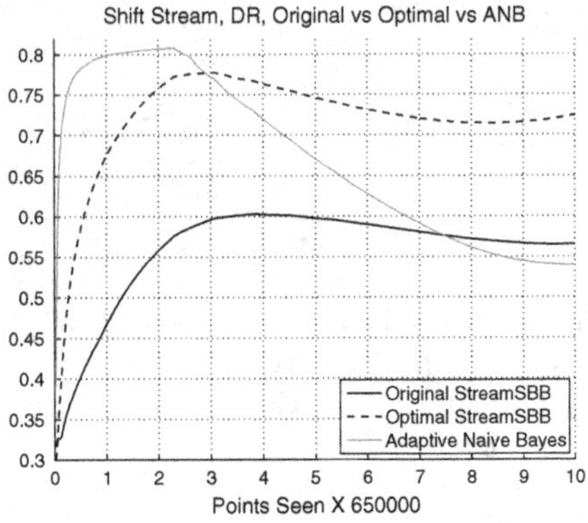

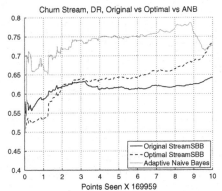

Fig. 8. DR curves for Concept Shift

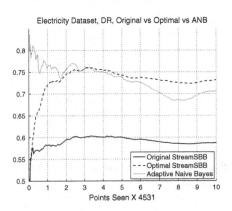

Fig. 9. DR curves for Electricity Demand

Fig. 10. DR curves for Churn Detection

ANB performance decays whereas StreamSBB performance generally continues to climb, ultimately resulting in StreamSBB reaching or exceeding the performance of ANB.

6   Conclusion

The StreamSBB framework has been revisited to support a TDL input representation. This is of fundamental importance when building classifiers for streaming applications. However, the dimensionality of the input space also undergoes a significant increase. Given that classifiers are built incrementally as the stream passes, it is necessary to make sure that the 'rate of evolution' is significantly higher than the rate at which the subset of labelled data 'turns over'. This is the role of DS oversampling ($\rho > 1$). Without this, StreamSBB performance is 10 to 15 % worse than originally configured. Moreover, this is performance as measured in terms of online class-wise detection rate, thus unaffected by merely improving performance on the majority class. This is the first time that such a metric has been demonstrated under empirical conditions (previous formulations being limited to offline evaluation scenarios).

Acknowledgments. The authors gratefully acknowledge support from NSERC Discovery and CRD programs (Canada) and RUAG Schweiz AG (Switzerland) while conducting this research.

References

1. Atwater, A., Heywood, M.I.: Benchmarking Pareto archiving heuristics in the presence of concept drift: diversity versus age. In: ACM Genetic and Evolutionary Computation Conference, pp. 885–892 (2013)
2. Atwater, A., Heywood, M.I., Zincir-Heywood, A.N.: GP under streaming data constraints: a case for Pareto archiving? In: ACM Genetic and Evolutionary Computation Conference, pp. 703–710 (2012)
3. Behdad, M., French, T.: Online learning classifiers in dynamic environments with incomplete feedback. In: IEEE Congress on Evolutionary Computation, pp. 1786–1793 (2013)
4. Bifet, A., Read, J., Žliobaitė, I., Pfahringer, B., Holmes, G.: Pitfalls in benchmarking data stream classification and how to avoid them. In: Blockeel, H., Kersting, K., Nijssen, S., Železný, F. (eds.) ECML PKDD 2013, Part I. LNCS, vol. 8188, pp. 465–479. Springer, Heidelberg (2013)
5. Cervantes, A., Isasi, P., Gagné, C., Parizeau, M.: Learning from non-stationary data using a growing network of prototypes. In: IEEE Congress on Evolutionary Computation, pp. 2634–2641 (2013)
6. Dempsey, I., O'Neill, M., Brabazon, A.: Foundations in Grammatical Evolution for Dynamic Environments. SCI, vol. 194. Springer, Heidelberg (2009)
7. Dempsey, I., O'Neill, M., Brabazon, A.: Survey of EC in dynamic environments (chap. 3). In: [6], pp. 25–54. Springer, Heidelberg (2009)
8. Doucette, J.A., McIntyre, A.R., Lichodzijewski, P., Heywood, M.I.: Symbiotic coevolutionary genetic programming: a benchmarking study under large attribute spaces. Genet. Program. Evolvable Mach. **13**(1), 71–101 (2012)
9. Fan, W., Huang, Y., Wang, H., Yu, P.S.: Active mining of data streams. In: Proceedings of SIAM International Conference on Data Mining, pp. 457–461 (2004)
10. Folino, G., Papuzzo, G.: Handling different categories of concept drifts in data streams using distributed GP. In: Esparcia-Alcázar, A.I., Ekárt, A., Silva, S., Dignum, S., Uyar, A.Ş. (eds.) EuroGP 2010. LNCS, vol. 6021, pp. 74–85. Springer, Heidelberg (2010)
11. Gama, J.: Knowledge Discovery from Data Streams. CRC Press, Boca Raton (2010)

12. Gama, J.: A survey on learning from data streams: current and future trends. Prog. Artif. Intell. **1**(1), 45–55 (2012)
13. Gama, J., Sebastião, R., Rodrigues, P.: On evaluating stream learning algorithms. Mach. Learn. **90**(3), 317–346 (2013)
14. Harries, M.: Splice-2 comparative evaluation: electricity pricing. Technical report, University of New South Wales (1999)
15. Heywood, M.I.: Evolutionary model building under streaming data for classification tasks: opportunities and challenges. Genet. Program. Evolvable Mach. (2015). doi:10.1007/s10710-014-9236-y
16. Lindstrom, P., MacNamee, B., Delany, S.J.: Drift detection using uncertainty distribution divergence. Evol. Intel. **4**(1), 13–25 (2013)
17. Polikar, R., Alippi, C.: Guest editorial: learning in non-stationary and evolving environments. IEEE Trans. Neural Netw. Learn. Syst. **25**(1), 1–3 (2014)
18. Žliobaitė, I., Bifet, A., Pfahringer, B., Holmes, G.: Active learning with drifting streaming data. IEEE Trans. Neural Netw. Learn. Syst. **25**(1), 27–54 (2014)
19. Žliobaitė, I., Gabrys, B.: Adaptive preprocessing for streaming data. IEEE Trans. Knowl. Data Eng. **26**(2), 309–321 (2014)
20. Vahdat, A., Atwater, A., McIntyre, A.R., Heywood, M.I.: On the application of GP to streaming data classification tasks with label budgets. In: ACM Genetic and Evolutionary Computation Conference: ECBDL Workshop, pp. 1287–1294 (2014)
21. Zhu, X., Zhang, P., Lin, X., Shi, Y.: Active learning from stream data using optimal weight classifier ensemble. IEEE Trans. Syst. Man Cybern. Part B **40**(6), 1607–1621 (2010)

Cartesian GP in Optimization of Combinational Circuits with Hundreds of Inputs and Thousands of Gates

Zdenek Vasicek[(⊠)]

Faculty of Information Technology, IT4Innovations Centre of Excellence,
Brno University of Technology, Brno, Czech Republic
vasicek@fit.vutbr.cz

Abstract. A new approach to the evolutionary optimization of large digital circuits is introduced in this paper. In contrast with evolutionary circuit design, the goal of the evolutionary circuit optimization is to minimize the number of gates (or other non-functional parameters) of already functional circuit. The method combines a circuit simulation with a formal verification in order to detect the functional inequivalence of the parent and its offspring. An extensive set of 100 benchmarks circuits is used to evaluate the performance of the method as well as the utilized evolutionary approach. Moreover, the role of neutral mutations in the context of evolutionary optimization is investigated. In average, the method enabled a 34 % reduction in gate count even if the optimizer was executed only for 15 min.

Keywords: Genetic programming · Cartesian Genetic Programming · Evolutionary optimization · Combinational circuits · Formal verification

1 Introduction

One of the most serious problems of evolvable hardware, especially in the area of evolutionary synthesis of logic circuits, is a very time consuming evaluation of candidate circuits. This problem is known as the problem of scalability. It causes that the evolutionary synthesis can handle only small and usually simple problems that are far from real-world problem instances.

In order to improve the scalability of evaluation, application-specific hardware as well as software methods were designed to increase the performance of the evolutionary optimization and design of logic circuits, see e.g. [2,4–6,9]. These methods enabled to increase the complexity of problem instances that can be solved in a reasonable time. Unfortunately, the methods are not scalable. The time needed to evaluate a candidate solution usually grows exponentially with the increasing number of primary inputs, but the accelerators are usually able to deliver a linear speedup only. Introducing more domain knowledge and utilizing more advanced evolutionary methods seem to be the only viable approach for dealing with the real-world problem instances. A breakthrough in the field of

© Springer International Publishing Switzerland 2015
P. Machado et al. (Eds.): EuroGP 2015, LNCS 9025, pp. 139–150, 2015.
DOI: 10.1007/978-3-319-16501-1_12

evolvable hardware was achieved with the introduction of a method which ties formal verification together with evolutionary optimization and substantially reduces the scalability issue of the evaluation [7]. Vasicek and Sekanina demonstrated that the previous empirical limitation of evolutionary design represented by a digital circuit having about 20 inputs can easily be overcome.

The goal of this paper is to introduce and evaluate a new approach which extends the method published in [8]. The advantage of the improved approach, which combines formal verification with simulation-based verification, is the ability to optimize digital circuits (i.e. to reduce the number of gates, improve power consumption, delay, etc.) represented at the gate level having hundreds of inputs and consisting of thousands of gates. The circuits of such a complexity have never been either evolved or optimized in the field of evolvable hardware at the gate level directly. In contrast with previously published works which are evaluated using a few benchmark circuits, an extensive set of 100 benchmarks circuits is used to evaluate the performance the proposed method. In addition to that, we would like to identify the key weaknesses of the evolutionary approach and propose future directions that could help the evolutionary approaches to penetrate into the area of real applications. In particular, we analyzed the role of neutral mutations in the context of evolutionary optimization.

2 Evolutionary Optimization of Combinational Circuits

2.1 Cartesian Genetic Programming

Cartesian Genetic Programming can be considered as one of the most efficient methods for evolutionary design and optimization of digital combinational circuits [3]. A candidate circuit is represented using an array of gates arranged in a matrix consisting of n_c columns and n_r rows. Each gate can be connected either to the output of a gate placed in previous l columns or to one of the circuit inputs. It means that no feedback is allowed. This requirement guarantees that only the combinational circuits will arise. Each gate is programmed to perform one of n_a-input functions defined in the set Γ. The number of circuit inputs, n_i, and outputs, n_o, is fixed. Every candidate circuit is encoded using $n_c \cdot n_r \cdot (n_a + 1) + n_o$ integers. The main advantage of the utilized encoding is that the size of phenotype is variable even if the size of chromosome is fixed. The variability is given by the fact that some nodes need not be employed in encoded circuit.

CGP operates with the population of $1+\lambda$ individuals. The initial population is usually seeded randomly. However, in order to optimize a known circuit (i.e. to minimize the number of gates), it is useful to seed the initial population by this circuit. Every new population consists of the best individual of the previous population and its λ offspring individuals. The offspring individuals are created using a point mutation operator which modifies h randomly selected genes of the chromosome. An important rule for selection of the new parent is utilized. In the case when two or more individuals can serve as the parent, an individual which has not served as the parent in the previous generation will be selected as a new parent. This strategy is important because it ensures the diversity of population [3]. The algorithm is terminated when the maximum number of generations is exhausted or a sufficient solution is obtained.

In case of digital circuit evolution, the fitness value of a candidate circuit is defined as follows. If a fully functional solution is evolved, the fitness value consist of the number of correct output bits obtained as response for all possible assignments to the inputs plus the number of unused CGP nodes. Otherwise, only the number of correct output bits is used. It means that the evolution has to discover a perfectly working solution firstly while the size of circuit is not important. Then, the number of gates is optimized. Similarly, delay or power consumption may be optimized.

2.2 Speeding up the Fitness Evaluation Using a SAT Solver

Contrary to the evolutionary design, the evolutionary optimization of digital circuits begins with the population seeded by a fully functional circuit. Usually, the goal is to minimize the number of gates. The most important feature of the evolutionary optimization is that each candidate solution created by means of genetic operators must be functionally equivalent with its parent in order to be further evaluated. This feature was utilized in [7] and furthermore elaborated in [8]. Equivalence checking was applied to decide whether a candidate circuit is functionally correct or not. In order to calculate the fitness value, the candidate circuit as well as its parent are converted to a Boolean formula whose satisfiability is investigated using a SAT solver. In fact, the parent serves as a golden reference for combinational equivalence checking. The advanced version, introduced in [8], utilizes another feature of evolutionary-based approach – the knowledge of the points in a candidate circuit that may break the correct function. This information is available because each offspring was created by a mutation from its parent. Hence, only a 'difference' (so-called cone of influence) between the candidate solution and its parent can be calculated. The Boolean formula can be derived from this 'difference'. Since the cone of influence usually represents only a small part of the candidate circuit, the time needed to decide the satisfiability of the Boolean formula can significantly be reduced.

If the obtained Boolean formula is satisfiable, a negative fitness value is assigned to the candidate circuit because the candidate circuit captures a different Boolean function. Otherwise, the candidate circuit is functionally equivalent with the specification and the fitness value is calculated according to the objective of the optimization. For example, the number of utilized gates was used in [7,8].

The usage of SAT solver helped to reduce the most time consuming part of the evolutionary algorithm, the evaluation of candidate solutions. In contrast with a common fitness function based on computing a Truth table, the time of evolution was reduced by several orders depending on the circuit parameters [8].

3   Proposed Method

In order to improve the performance of the evolutionary optimizer, i.e. to increase the number of candidate solutions that can be evaluated within a period of time, we suggest to combine SAT solver with a circuit simulator which will be used to disprove the equivalence between a candidate solution and its parent. This approach is based on the assumption that the time needed to simulate a given candidate circuit using N_V ($N_V \ll 2^{n_i}$) test vectors (t_{sim}) is significantly lower than the time which is consumed by a SAT solver (t_{sat}).

The correctness of a candidate solution is determined as follows. Firstly, a circuit simulator is applied to the difference circuit between a candidate solution and its parent (difference circuit is calculated according to [8]). The simulator can use up to N_V randomly generated test vectors. If there is a test vector which evaluates the output of the difference to one, the simulator is terminated and a negative fitness value is assigned to the corresponding candidate solution. Since it is guaranteed that the candidate solution is not functionally equivalent with its parent, it is not necessary to call SAT solver to prove that fact. Otherwise, when all N_V test vectors are applied and the output of the difference evaluates to zero in all the cases, a SAT solver has to be used to prove or disprove the equivalence because the limited number of test vectors cannot guarantee that there is not a vector that differentiates the circuits.

The speedup of the proposed method combining a simulator and SAT solver can be defined as follows:

$$gain = \frac{t_{sat}}{t_{sim} + \sigma_{fail}t_{sat}} = \frac{1}{t_{sim}/t_{sat} + \sigma_{fail}}, \tag{1}$$

where $\sigma_{fail} = [0, 1]$ is a coefficient which determines the fail-rate of the simulation-based equivalence checking. The σ_{fail} may also be understood as the probability of occurrence of an undetected fault.

If we want to maximize the gain, i.e. the overall performance of the optimizer, we need to minimize not only the value of the ratio t_{sim}/t_{sat}, but also the value of σ_{fail}. Even if the simulator is e.g. 1000 times faster than SAT solver, a negligible improvement will be achieved if the value of σ_{fail} is close to one. The value of t_{sim} as well as σ_{fail} depend on the number of test vectors that can be used in the simulator to disprove the equivalence. While t_{sim} increases linearly with increasing N_V and the size of the difference entering the simulator, σ_{fail} decreases with increasing N_V. Hence, appropriate value of N_V has to be determined in order to maximize the gain.

4   Experimental Results

4.1   Benchmark Circuits

In order to evaluate the performance of the proposed method, we utilized a set of 100 randomly chosen circuits form QUIP, WLSI and ACM/SIGDA benchmark set (only circuits with 15 and more primary inputs are considered). These circuits were synthesized and optimized by ABC[1] using 'choice' script. The result of ABC was utilized as the input to the evolutionary optimizer.

---

[1] ABC is a system for sequential synthesis and verification by A. Mishchenko.

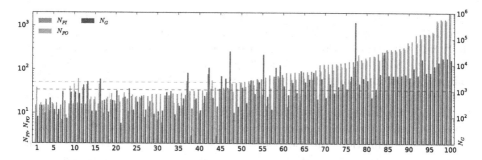

Fig. 1. The number of primary inputs (N_{PI}), primary outputs (N_{PO}) and gates (N_G, right axis) for each benchmark circuit. The X-axis contains the index of benchmark circuit. The benchmarks are arranged according to the increasing complexity expressed as $2^{N_{PI}}N_G$. Note that both Y-axes have a logarithmic scale (The list of benchmark circuits is available at http://www.fit.vutbr.cz/~vasicek/gp15).

The basic parameters of the benchmark circuits are given in Fig. 1. The circuits are arranged according to the increasing complexity. The complexity is expressed as a time needed to evaluate a candidate solution using a common fitness function (i.e. the fitness function based on a truth table). In such a case, the evaluation time is dependent on two factors: the number of primary inputs (N_{PI}), and the number of gates (N_G). As the time needed to evaluate a candidate solution increases exponentially with the increasing number of primary inputs, N_{PI} represents the key parameter which has a great impact on the total time.

The least complex circuit, 'alcom' circuit with index 1, consists of 106 gates and utilizes 15 primary inputs and 38 outputs. The most complex circuit, audio codec controller 'ac97_ctrl' with index 100, contains 16158 gates and uses 2176 inputs and 2136 outputs. One half of the benchmark circuits have more than 50 primary inputs and consist of more than 1000 gates.

4.2   Role of Neutral Mutations

The objective of the first experiment was to confirm or reject hypothesis about the importance of neutral mutations in evolutionary optimization of combinational circuits. Two variants of the mutation operator were implemented in order to evaluate the significance of neutrality. The first implementation does not impose any special limitations on the mutation operator. The only requirement is to modify the value of a randomly chosen gene to a different one (but legal). On the other hand, three restrictions are applied in the second implementation: (1) inactive gates are never modified; (2) it is not possible to connect an active gate (or primary output) to an inactive gate; (3) the gene which encodes the connection of the second input of a single-input gate is never mutated. These restrictions were introduced in order to mitigate the neutral mutations.

The CGP parameters were chosen as follows: $n_c = N_G$, $n_r = 1$, $l = N_G$, $\lambda = 1$, $h = 2$, $\Gamma = \{\text{BUF, INV, AND, OR, XOR, NAND, NOR, XNOR}\}$. These parameters were chosen according to the [8]. No redundancy in CGP encoding is used; the number of nodes is equal to the size of a benchmark circuit obtained from ABC. The goal of CGP is to minimize the number of utilized gates, i.e. the fitness value is equal to the number of active CGP nodes. The fitness function utilizes SAT solver only. In order to perform a statistical evaluation, fifteen independent evolutionary runs were executed for each benchmark circuit. Note that median value will be used to analyze the impact of a particular parameter because no Gaussian distribution can be observed among the benchmarks. The evolution is terminated after $15\,\mathrm{min}^2$. We do not use the number of evaluations as a termination condition because this number is very sensitive to the structural properties of an optimized circuit and it is impossible to determine an appropriate value in advance.

The performance of both approaches is evaluated using the number of generations (G_{impr}) that enabled an improvement of the fitness value. This parameter can be seen as a measure of mutation operator's performance (i.e. the ability to generate a candidate solution which is valid and simultaneously improved). The reason behind the usage of this metric is that the number of evaluations cannot be compared directly because the neutral mutations are detected and the created candidate solutions do not enter the time-consuming fitness evaluation procedure (it is guaranteed that they have the same fitness value as their parent) resulting in the fact that significantly more generations can be produced if the occurrence of neutral mutations is high.

Let $G = G_{valid} + G_{invalid}$ be the total number of generations where G_{valid} is the number of generations in which a valid candidate solution (i.e. functionally equivalent with a parental circuit) is generated from a parental solution by applying the mutation operator. Then, G_{valid} can be expressed as $G_{valid} = G_{impr} + G_{noimpr} + G_{neutral}$, where $G_{neutral}$ is the number of neutral mutations in the sense defined in previous paragraphs. G_{noimpr} represents the candidate solutions in which at least a single gene was changed but the fitness value remained unchanged. Note that $G_{neutral} = 0$ in the second implementation because no neutral mutations are allowed.

The evaluation of both variants of the mutation operator is shown in Fig. 2. The performance is expressed as the ratio $G_{impr}/(G_{valid} - G_{neutral})$ calculated at the end of each 15-minute evolutionary run, averaged over all fifteen runs. Despite the stochastic nature of evolutionary algorithm which leads to some variances (see the error bars in Fig. 2 showing the magnitude of standard deviation), we can conclude that the performance of both implementations is almost identical. In average, 2.34 % of valid generations were produced when the neutral mutations were enabled and 2.42 % for the opposite case. For 75 benchmarks, the variant with disabled neutral mutations performs approx. $(30 \pm 35)\,\%$ better in average. The performance was worsened in 25 cases by approx. $(9 \pm 10)\,\%$ in average.

According to the obtained results, we can conclude that it has no advantage to support neutral mutations in this scenario (i.e. if the goal is to minimize the

---

[2] A PC equipped with Intel Xeon X5670 (24 cores, 2.93 GHz, 12 MB cache), 32 GB RAM and 64-bit CentOS Linux was used.

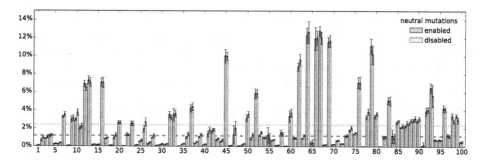

Fig. 2. The mean number of generations that enabled an improvement of fitness value when the neutral mutations were enabled (disabled). It is expressed as the ratio $G_{impr}/(G_{valid} - G_{neutral})$. The mean value obtained as an average over all benchmarks is represented by dotted line whereas the median value is depicted by dash-line.

number of gates in a fully functional circuit). In fact, the neutral mutations have a negative impact on overall performance because the probability of mutation of an active gene decreases with the increasing number of inactive genes. Even if the neutral mutations are detected and the corresponding candidate solutions do not enter the time-consuming fitness evaluation procedure, the performance of the evolutionary optimizer deteriorates as the circuit is reduced because a great portion of neutral mutations is generated.

Looking at the results shown in Fig. 2, we can identify that the performance of the mutation operator is very sensitive to the optimized circuit. One can admit that this issue could be related to the impossibility to improve the number of gates of a given benchmark circuit, but this is certainly not the case. It can be easily shown that the utilized circuits are not optimal if the number of gates is considered. Taking into account that the ratio between G_{valid} and G is approx. 0.5 % in average (see Fig. 3), there are circuits for which the mutation

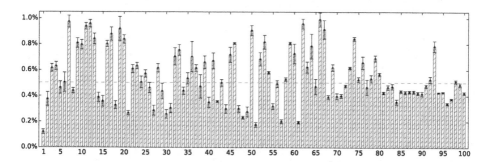

Fig. 3. The number of generations in which a valid candidate solution was produced, represented as a ratio $G_{valid}/(G_{valid} + G_{invalid})$. The results are obtained from the second implementation, where the neutral mutations are disabled. The median value is shown using a dash-line.

operator performs very poorly. Less than 0.007 % of the total number of genera-tions enabled the improvement of the fitness value for one half of the benchmark circuits. On the other hand, there are instances showing a significantly better convergence, e.g. more than 0.12 % of the total number of generations leading to the improvement of the fitness value were produced in the case of circuit 66.

Unfortunately, there is no obvious relation between the circuit complexity (as defined in Sect. 4.1) and performance of the mutation operator. Thus, we believe that the performance of the mutation operator is in a close relation with the internal structure of an optimized circuit. Hence there are two possibilities how to improve the performance of the evolutionary optimizer. We can (a) increase the number of generations that can be evaluated within a time period and/or (b) to design a new mutation operator with better performance.

4.3   Efficiency of the Proposed Approach

To determine the value of σ_{fail} and its dependency on N_V, three experiments were performed. A 64-bit parallel simulator which is able to calculate response to 64 input combinations in a single pass was utilized. The simulator was enabled to use (a) a single pass ($N_V = 64$), (b) up to 16 passes ($N_V = 1024$), and (c) up to 32 passes ($N_V = 2048$) to disprove the equivalence. Only the cone of influence determined according to the points of mutation enters the simulator. The experimental setup and CGP parameters were the same as described in previous section. The mutation operator with suppressed neutral mutations was employed.

The obtained results are shown in Fig. 4. The value of σ_{fail} was calculated at the end of fifteen 15 min evolutionary runs. The median value of N_V can be approximated by the exponential trendline $\overline{\sigma}_{fail} \approx 3.2693 N_V^{-0.611}$ with R-squared equal to 0.9955. It means that σ_{fail} noticeably decreases at the begin-ning (i.e. for small N_V) and then, as N_V increases, the yield is smaller and smaller. In most cases, σ_{fail} is lower than 0.1 even if a single pass is used. How-ever, there are cases with surprisingly high ratio of σ_{fail} that remains above 50 %

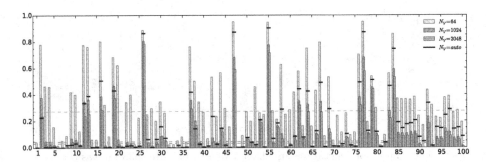

Fig. 4. Fail-rate σ_{fail} of simulation-based equivalence checking shown for various num-ber of randomly generated test vectors (N_V) that are utilized by the circuit simulator to disprove functional equivalence between candidate solution and its parent.

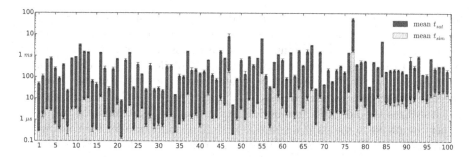

Fig. 5. Average time needed to perform equivalence checking using (a) SAT solver (see t_{sat}) and (b) simulator with a single pass (see t_{sim}).

even if 2048 randomly generated input combinations were utilized (see benchmarks 26, 47, 55, 77 and 84). Considering the parameters of those circuits (see Fig. 1), we suppose that this issue is probably related to the high number of utilized gates which may contribute to a fault masking effect.

The σ_{fail} corresponding to the number of test vectors that are used to maximize value of Eq. 1 is represented by lines labeled as $N_V = auto$ in Fig. 4. We can observe that less than 16 passes (i.e. less than 1024 test vectors) were used in most cases. These instances can easily be identified by comparing the value of σ_{fail} for $N_V = auto$ and $N_V = 1024$; the lower number of test vector implies higher σ_{fail}. Unfortunately, the ratio t_{sat}/t_{sim} remains very low for the five benchmarks discussed in previous paragraph (see Fig. 5). Hence only a few test vectors can be utilized which results in the fact that the fail-rate remains very high. Thus only a negligible speedup is expected in these cases.

The speedup of the proposed method combining SAT solver with simulator is given in Fig. 6. The speedup is calculated using the number of candidate solutions that can be evaluated within 15 min. The number of test vectors was determined adaptively during the evolution as follows. At the beginning of the evolution, a

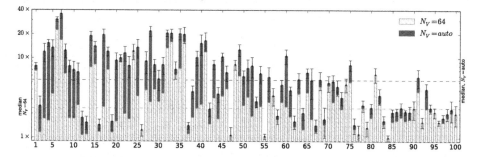

Fig. 6. Speedup of the proposed method which combines SAT-based and simulation-based equivalence checking in the fitness function. For more than 50 benchmark circuits, adaptive setting of the number of test vectors (see $N_V = auto$) increased the speedup approx. twice compared to a single-pass simulation (i.e. 64 test vectors). Note that the y-axis has a logarithmic scale.

single pass (i.e. 64 test vectors) is utilized. Then, the number of passes doubles every 10 s until a decrease in the performance is detected. Finally, the best value is determined and used. The number of test vectors is adaptively modified during evolution if there exists a different value which provides better performance.

According to the obtained results, the achieved speedup is higher than 5.28 for half of the benchmark circuits. The performance of the implementation which utilizes the adaptive number of test vectors is approximately two times higher compared to the implementation with fixed number of test vectors whose speedup factor is approx. 2.34. This finding can be considered as a very positive result since the introduction of the simulator can remarkably improve the performance of the evolutionary optimizer.

Similarly to our previous findings regarding σ_{fail}, the value of speedup noticeably varies across the benchmarks. There are cases for which the speedup factor exceeded 30. On the other hand, nearly no improvement was obtained for benchmarks 26, 47, 55, 77, and 84. According to our expectation, the speedup is close to 1.0 in these cases.

We analyzed the obtained results and identified that there is a relation between σ_{fail} and speedup. If $\sigma_{fail} \geq 0.05$, the higher σ_{fail} implies a lower speedup. However, this relation does not hold for $\sigma_{fail} < 0.05$ where the speedup varies in one order independently on the value of σ_{fail}. In addition to that, we can observe decreasing of the t_{sat}/t_{sim} ratio as the complexity of a benchmark circuit increases. Even if t_{sat} remains relative stable across the benchmarks (see Fig. 5), t_{sim} increases with the increasing complexity. The ratio t_{sat}/t_{sim} was decreasing from approx. 350 for less complex circuits to 10 for the most complex circuit. As a consequence of that, a relative small number of test vectors should be used in the simulator.

4.4   Performance of the Circuit Optimizer

The impact of the proposed method on the quality of optimization is shown in Fig. 7. The implementation which utilizes SAT solver and circuit simulator with

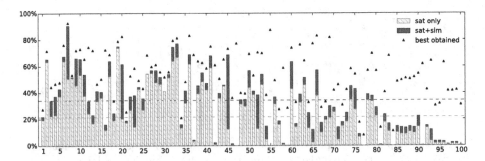

Fig. 7. Reduction of the benchmark circuits (relative to the original size) obtained after 15 min of the optimization is shown for (a) sat-based optimizer and (b) the proposed approach which combines SAT solver with simulator. The best results obtained from a 24-hour evolutionary optimization are denoted by triangles.

adaptive number of test vectors is compared against the SAT-based implementation introduced in [8]. No neutral mutations were enabled. The experimental setup is the same as used in previous section.

In all cases, the combination of a SAT solver and circuit simulator brought an improvement. The size was reduced by 13 % in average. Still, there are cases showing a very slow convergence caused mainly by the time consuming evaluation. If we compare average G_{valid} of the four aforementioned benchmarks (26, 47, 55, 77, and 84) with G_{valid} of the rest of the benchmarks, we can observe that the value is two orders of a magnitude lower. This explains why nearly no improvement was achieved within 15 min in these cases.

5   Conclusion

We introduced a new approach to the evolutionary optimization of large digital circuits which exploits the combination of a circuit simulator and a formal verification. Due to the usage of a simulator with adaptive number of test vectors, the time of evaluation was significantly reduced for 100 complex benchmark circuits in comparison with a method published in [8]. In the worst case, the time of evaluation remains the same.

In addition to that, we investigated the role of neutral mutations that are believed to be an important part of CGP. According to the obtained results, we have concluded that it has no advantage to support neutral mutations for circuit optimization (i.e. in the case that the number of gates is minimized for a fully functional circuit). This can be understood as an important result not only from theoretical but also from practical point of view because the neutral mutations in fact have negative impact on the performance of the evolutionary optimization. Our findings related to the role of neutrality correspond with observations on the evolutionary design of parity circuits [1].

The performance of the proposed method was evaluated on an extensive set of real-world benchmark circuits having tens to hundreds of inputs and consisting of hundreds to thousands of gates. For more than half of the benchmark circuits, approximately five times higher number of evaluations was performed within the same time period compared to the approach that utilizes only a formal approach. While the latter method was able to reduce the circuits by 21 % in average, the proposed method is able to reduce the circuits by 34 % using the same amount of time. Considering the fact that the runtime of the optimization process was 15 min, the obtained results are very encouraging.

We demonstrated that the circuit optimization conducted by CGP is applicable on complex real-world digital circuits. However, we simultaneously shown that there are instances for which the proposed method can bring only a marginal or none improvement in the performance. Our method is based on the assumption that evolutionary-based approach generates a large number of invalid candidate solutions that can be detected very quickly by means of applying a few test vectors on the inputs (i.e. that the time consuming formal verification can be replaced with a faster simulation-based approach). While this assumption is valid and an enormous number of invalid candidate solutions are generated during evolution, there exist circuits that are hard for the simulation-based verification. We believe that the evolutionary-based approach requires to generate a large number of candidate solutions to compensate the poor performance of the mutation operator. We observed that at least $5 \cdot 10^4$ valid candidate solutions were generated within 15 min for problem instances exhibiting a reasonable convergence. Unfortunately, approx. two orders of a magnitude (i.e. 10^6) candidate solutions have to be generated to obtain $5 \cdot 10^4$ valid candidate solutions.

One of the possibilities how to substantially improve performance of the evolutionary optimization is to orient the future research towards improving of the mutation's operator performance. Another option is to replace the randomly generated test vectors with a smart selection of test vectors which can quickly detect the inequivalence. One of the possibilities is to build a database of test vectors using the counter examples that are produced by a SAT solver during verification.

Acknowledgments. This work was supported by the Czech science foundation project 14-04197S.

References

1. Collins, M.: Finding needles in haystacks is harder with neutrality. Genet. Program. Evolvable Mach. **7**(2), 131–144 (2006)
2. Harding, S., Miller, J.F., Banzhaf, W.: Self modifying Cartesian genetic programming: parity. In: 2009 IEEE Congress on Evolutionary Computation, pp. 285–292. IEEE Press (2009)
3. Miller, J.F.: Cartesian Genetic Programming. Springer, Heidelberg (2011)
4. Shanthi, A.P., Parthasarathi, R.: Practical and scalable evolution of digital circuits. Appl. Soft Comput. **9**(2), 618–624 (2009)
5. Stomeo, E., Kalganova, T., Lambert, C.: Generalized disjunction decomposition for evolvable hardware. IEEE Trans. Syst. Man Cybern. Part B **36**(5), 1024–1043 (2006)
6. Vasicek, Z., Sekanina, L.: Hardware accelerators for Cartesian genetic programming. In: O'Neill, M., Vanneschi, L., Gustafson, S., Esparcia Alcázar, A.I., De Falco, I., Della Cioppa, A., Tarantino, E. (eds.) EuroGP 2008. LNCS, vol. 4971, pp. 230–241. Springer, Heidelberg (2008)
7. Vasicek, Z., Sekanina, L.: Formal verification of candidate solutions for postsynthesis evolutionary optimization in evolvable hardware. Genet. Program. Evolvable Mach. **12**(3), 305–327 (2011)
8. Vasicek, Z., Sekanina, L.: A global postsynthesis optimization method for combinational circuits. In: Proceedings of the Design, Automation and Test in Europe, DATE, pp. 1525–1528. IEEE Computer Society (2011)
9. Walker, J.A., Miller, J.F.: The automatic acquisition, evolution and re-use of modules in Cartesian genetic programming. IEEE Trans. Evol. Comput. **12**(4), 397–417 (2008)

Posters

Genetic Programming for Feature Selection and Question-Answer Ranking in IBM Watson

Urvesh Bhowan$^{(\boxtimes)}$ and D.J. McCloskey

IBM Ireland (IBM Technology Campus), Dublin, Ireland
{urvesh.bhowan,dj_mccloskey}@ie.ibm.com

Abstract. IBM Watson is an intelligent open-domain question answering system capable of finding correct answers to natural language questions in real-time. Watson uses machine learning over a large heterogeneous feature set derived from many distinct natural language processing algorithms to identify correct answers. This paper develops a Genetic Programming (GP) approach for feature selection in Watson by evolving ranking functions to order candidate answers generated in Watson. We leverage GP's automatic feature selection mechanisms to identify Watson's key features through the learning process. Our experiments show that GP can evolve relatively simple ranking functions that use much fewer features from the original Watson feature set to achieve comparable performances to Watson. This methodology can aid Watson implementers to better identify key components in an otherwise large and complex system for development, troubleshooting, and/or customer or domain-specific enhancements.

Keywords: Genetic Programming · IBM Watson · Question answer ranking · Feature selection

1 Introduction

IBM Watson is an intelligent open-domain question answering (QA) system capable of answering questions posed in rich natural language in real-time [7]. The open-domain QA problem is one of the most challenging in computer science and artificial intelligence as it leverages aspects from information retrieval (IR), natural language processing (NLP), knowledge representation, machine learning (ML) and complex reasoning. Watson gained international attention after beating human champions on the American TV quiz show, *Jeopardy!* [7]. Since Jeopardy!, Watson has shown success in many other commercial domains such as health care, finance and customer engagement [8].

IBM Watson uses the *DeepQA* architecture, a massively parallel probabilistic and evidence-based approach, to search and reason over large volumes of unstructured information [9]. DeepQA uses ML to rank candidate answers generated by the system in response to an input question using a large extremely heterogeneous feature set derived from many distinct and independently developed NLP and IR algorithms [10]. However, there is need by Watson developers

© Springer International Publishing Switzerland 2015
P. Machado et al. (Eds.): EuroGP 2015, LNCS 9025, pp. 153–166, 2015.
DOI: 10.1007/978-3-319-16501-1_13

and implementers to better understand the feature contributions in the data, in particular, to find the most useful features in the system for a given task or domain. This knowledge can aid Watson implementers identify key components for troubleshooting and/or delivering faster customer or domain-specific enhancements in an otherwise large and complex system.

Genetic Programming (GP), an evolutionary ML algorithm, has shown success in feature selection [4,12] and in the IR domain to automatically learn numeric ranking functions to order relevant or non-relevant web documents [16,18]. Learning IR ranking functions is difficult as these are typically manually designed by experts based on heuristics and statistical theories. However, feature selection is not a major requirement in these works where, unlike Watson, these web ranking tasks typically use small, carefully selected (by domain experts) and relatively homogeneous feature sets [16,17].

This paper attempts to bridge these two domains (GP feature selection and question-answer ranking). Our main goal is to develop a two-phase domain-independent GP approach for feature selection in IBM's NLP question answering system, Watson. In the first phase, we leverage GP's automatic feature selection mechanisms to evolve *simple* (small) but highly accurate functions to rank and classify candidate answers generated by Watson. Here we limit the evolved tree sizes to increase selection pressure for good features during evolution. In the second phase, we leverage GP's model transparency/interpretability properties to mine the pool of evolved GP trees of varying complexity/size (from the first phase) to automatically extract feature subsets; these represent key features automatically identified through the learning process. We evaluate the proposed GP approach on English general knowledge questions with factoid answers in English Wikipedia. Our experiments involve two main investigations. The first evaluates the performance of the evolved GP functions to directly rank and classify answer answers generated by Watson. The second evaluates Watson's performance (using Watson's ML framework) using feature subsets automatically generated from the pool of evolved GP trees. Our experiments find that GP can successfully identify very small feature subsets (using fewer than 8 % of all Watson's total features) that perform to within 90 % Watson's overall accuracy.

The rest of this paper is identified as follows. Section 2 introduces relevant background in Watson and related work in GP. Section 3 outlines the GP approach for question answer ranking. Section 4 presents the experimental results which applies GP directly to the answer ranking task. Section 5 presents our methodology for building our GP feature subsets and presents the feature selection results. Section 6 presents our conclusions and future work.

2   Background and Related Work

This section introduces Watson in more detail, presents related work in GP for feature selection and answer ranking, and outlines the main challenges for our approach.

2.1   Watson Overview

In 2011, Watson gained international attention when it beat two human champions on the American TV quiz show, *Jeopardy!* [7]. Here factoid-type general knowledge questions over are posed to competitors in rich natural language, where the first to buzz in with a correct answer wins points. Watson uses the *DeepQA* architecture, a massively parallel, probabilistic and evidence-based approach, to answer natural language questions in real-time [9]. The DeepQA pipeline uses four main phases to answer a question. **Question Analysis** performs a detailed syntactic and semantic analysis of the input question using various NLP technologies such as a natural language parser and named entity recognizers. **Hypothesis Generation** builds candidate answers to a question by searching over a corpora (such as Wikipedia, as used in this paper) and extracting potential answers from the search hits. **Hypothesis Scoring** uses many NLP algorithms to score the relevance of candidate answers to the question. Each scoring algorithm outputs one or more features that measure how well the evidence supports a candidate answer (the Watson configuration in this paper uses 354 features). In **Final Merging and Ranking** (FMR), similar candidate answers are merged and ML is used to rank the merged set of answers based on their feature scores. Here the top-ranked candidate is selected as the final answer to a given question.

The ML phase in Watson estimates the probability that a generated candidate answer (in response to a question) is correct [10]. A cascade of (binary) logistic regression classifiers is employed in successive phases, where all outputs from one phase are passed as input to the classifier in the next phase. Once the probabilities of all candidate answers for a given question are obtained, the candidate with the highest probability is selected as the final answer to a given question. Each logistic regression classifier is trained (offline) using existing questions and their correct answers.

2.2   Related Work in GP

Feature Selection. Watson generates a large feature set (354 features) in its ML phase. Within IBM Watson, Pearson's Coefficient and Gram-Schmidt orthonormalising (GSO) have been used for feature analysis to estimate feature informative in the data [10]. Both are deterministic numerical analysis methods to rank features in order of decreasing relevance to the target output (class label). However, both offer naive and "shallow" feature analysis as they only consider a linear combination of features. Machine learning algorithms such as Genetic programming (GP) can address this limitation by formulating feature selection as a combinatorial optimization problem. GP has been widely used for feature selection in the classification domain for two main tasks 2, mainly due to the implicit feature selection and construction mechanisms inherent in the evolution (see Sect. 3.3 for details on these mechanisms).

The first task is to find useful feature subsets from the original set (that maximizes some criterion) [1,3]. In [3], several GP classifiers are evolved (using the full feature set) and the most frequently used features from these GP trees are extracted; a second classifier is then evolved using only the extracted features. In [1], only feature terminals are used in the GP trees and tree depth is strictly limited to increase selection pressure for features in the evolved trees. Our work combines aspects from both these approaches: we limit GP tree depth for feature selection and extract frequently used features from the evolved GP ranking functions. The second task is to construct new composite features (via arithmetic operators on the original features) [13,15]. GP approaches for feature construction are split into filter or wrapper-based approaches. The former constructs features before classifier induction (e.g. [13]), whereas the latter interweaves feature construction and classifier induction (e.g. [15]). In both cases, the evolved GP trees represent new constructed features.

Answer Ranking. GP has typically been used to learn numerical ranking functions to order web documents from user queries, where documents are either is relevant or non-relevant (two classes) [5,6,16–18]. These approaches all use a tree-based GP representation, a relatively similar function/terminal set (to evolve numeric expressions), and a fixed tree depth to address bloat. All also use a ranking-based IR measure (such as Mean Average Precision) directly in the fitness function. By contrast, older work in this area tend to approach this problem in two-steps: solve as a classification problem and then use loss functions to train a ranking model. These works typically compare the evolved GP trees (on a given data set) to well-known (static) ranking functions manually designed by experts (such as Okapi-BM25). All emphasize that GP can perform as well as, or better than, the established manually designed functions.

In [5], GP outperforms Okapi-BM25 on news-wire document ranking from the Associate Press. In [17], multiple evolved GP functions and Okapi-BM25 are aggregated into a logistic regression function (as composite features) for improved accuracy, where this aggregate is shown to outperform individual components. In [6], GP functions are evolved for individual user search queries which outperform two established ranking functions (Okapi-BM25 and PTFIDF). In [16], GP is shown to evolve accurate general-purpose ranking functions across several unstructured IR tasks from the TREC (Text Retrieval Evaluation Conference) collection.

The GP representation in our paper is similar to these works except for two main differences. We use an extra conditional operator (if) for better program flexibility, and a finer-grain ranking measure in the fitness function (Mean Reciprocal Rank).

Major Challenges in Question-Answer Ranking. A major limitation in the related works are the comparatively small number of features considered in a given problem. Most use roughly 10 features [5,6,17], while other use 21 [16] or 40 [18] features. This means that feature selection is not a major requirement in these works. As a result, GP for feature selection in this domain has previously been explored (to our knowledge). By comparison, Watson uses 354 features – a considerably larger feature-space for the GP evolutionary search. Furthermore, the feature in the related works above tend to be relatively homogeneous. These are typically classic IR document/corpus statistics (such as term frequency, inverse document frequency, document length, etc.), based on tried-and-tested *a priori* domain knowledge in web document ranking. In contrast, Watson's features are extremely heterogeneous as they are derived from a variety of distinct and independently developed NLP algorithms. Features may be also radically different between questions (e.g. depending on the type and/or structure of the question). Another challenge is the large class imbalance inherent in the data. There are potentially hundreds of incorrect candidate answers generated by the system in response to a question compared only one or two correct answers.

3   GP Approach

This section outlines our GP representation, fitness function for question answer ranking, GP mechanisms for feature selection and other evolutionary parameters.

3.1   Evolving GP Functions for Ranking

For a given input question (e.g. "What is the capital city of Ireland?"), each candidate answer generated by Watson has a string label, an associated feature vector, and its relevance judgment (class label) [10]. The class label (`correct` or `incorrect`) is determined by comparing the candidate answer label to the ground-truth[1] containing correct answers for each question in a question set. General knowledge questions with factoid answers are considered, e.g., "**Dublin**", "Cork" or "Belfast" represent (factoid) candidate answers generated by Watson in response to the above question (where the answer(s) judged as `correct` by the ground-truth is highlighted in bold text).

The goal of an evolved GP function is to rank the candidate answers generated from the Watson system for each input question. Once all candidate answers for a question are ranked, the candidate in the first position is taken as the final answer. This ranking is based on the raw output of the evolved GP expression when the expression is evaluated on a candidate answer feature vector. The question is considered correctly answered if the top-ranked candidate answer according to the ranking function (in this GP, an evolved GP tree) has the `correct` class label. Otherwise, if the top-ranked candidate answer has the `incorrect`, the question is not answered correctly.

A tree-based structure is used to represent the evolved genetic programs [12]. Each GP solution represents a mathematical expression that outputs a (floating point) number when evaluated on a given input (candidate answer feature vector to be ranked). We use feature terminals (features) and constant terminals (randomly generated floating point numbers), and a function set comprising of the

---

[1] The ground-truth dictionary is manually created and curated by the Watson development team.

four standard arithmetic operators, $+$, $-$, $\times$, and $\%$, and the conditional operator if. The $+$, $-$ and $\times$ operators have their usual meanings (addition, subtraction and multiplication) while $\%$ is *protected* division (usual division except that a divide by zero returns zero).

The conditional if function takes three arguments and returns either the second argument if the first is negative, or the third argument otherwise. This function allows programs to contain expressions in different regions of feature-space and allows discontinuities rather than insisting on smooth functions. As mentioned (in Sect. 2.2), related works have not previously used a conditional if. This operator has been shown to improve program flexibility in classifiers [2].

3.2   Fitness Function for Answer Ranking

A statistical measure of rank, the Mean Reciprocal Rank (MRR), is used as the GP fitness function. The reciprocal rank of a question is the multiplicative inverse of the rank of the first correct answer (answer with correct class). For a set of questions Q, the MRR is the average of the reciprocal ranks, as shown in Eq. (1).

$$mrr = \frac{1}{|Q|} \sum_{i=1}^{|Q|} \frac{1}{\text{reciprocal rank}_i} \tag{1}$$

where

$$\text{reciprocal rank}_i = \arg\max[\text{GP}(a_{i,0}), \text{GP}(a_{i,1}), ..., \text{GP}(a_{i,m})]$$

In the equation above, $\text{GP}(a_{i,j})$ is the output of an evolved GP expression when evaluated on the jth candidate answer a for question i, where m is the number of candidate answers for question i. For example, the reciprocal rank will be 2 for the ordered list [Belfast, **Dublin**, Cork] – as the correct answer is in position two. The reciprocal rank will be 0 if no correct answers are generated for a question. If multiple correct answers occur in the list, the highest reciprocal rank is used.

The question accuracy (number of questions correctly answered) measure metric can also be used in GP as the fitness function. We chose the MRR for the fitness function as it is much more *fine-grained* measure than question accuracy. The MRR reflects subtle changes in the ranking across all candidate answers; whereas question accuracy only takes into account the top-ranked candidate answers, ignoring the ranked positions of the other candidates. For example, give the two following two ranked lists:

list$_1$ = [Belfast, Cork, **Dublin**] (reciprocal rank is 3)
list$_2$ = [Belfast, **Dublin**, Cork] (reciprocal rank is 2, so ranking is better).

Here the question accuracy for both lists are zero as the top-ranked answer in both lists are incorrect. However, the MRR judges that list$_2$ is better-ranked than list$_1$ as its reciprocal rank is higher.

3.3   GP Tree Size and Other Evolutionary Parameters

By imposing a strict limit on the maximum tree depth we aim to leverage GP's implicit feature selection and construction properties (due to stronger selection pressures for good features) in the evolution [4]. Since GP trees have variable lengths, not all features can be instantiated as a tree node in an individual when the tree depth is limited to relatively small GP trees. Implicit feature selection is performed through selection pressures for available nodes in a tree, as only a limited number of features can be instantiated in a individual during the evolution. Implicit feature construction is performed in a similar manner by the application of arithmetic operations in the trees (to process features). Using a limit on GP tree depth also helps to address bloat and overfitting. Note that the evolved trees do not have to be fully-formed.

In the experiments, we explore six maximum GP tree depth settings (between 4 and 9) where the maximum number of features used in an evolved GP tree depends on the maximum tree depth. The maximum number of leaf nodes (potential feature nodes) in a fully-formed GP tree is Ar^D where Ar is the highest arity in the GP function set (e.g. 3 due to if operator) and D is the maximum tree depth. Similarly, the maximum number of nodes in a fully-formed GP tree is $(Ar^{D+1} - 1)$. However, these represent generous estimates as it is unlikely that the evolved GP trees will be fully-formed, have all internal tree nodes set to the if operator, and have all leaf nodes set to feature terminals (no constant terminals). In practice, Ar is closer to 2 as the other arithmetic operators take two arguments (this can be seen in our experiments in Table 1 in the next section). Using this as a guide, the maximum number of feature nodes in the evolved GP trees can be between 16 and 512 for depths 4 and 9, respectively (but this estimate is also generous due to the same reasons previously discussed). As the Watson system has 354 total features, a maximum tree depth of 9 will allow many different feature combinations to be used in a single GP tree.

For the other evolutionary parameters, the ramped half-and-half method is used for generating genetic programs in the initial population and for the mutation operator [12]. The population size is 500. The crossover, mutation and elitism rates are 60 %, 35 % and 5 %, respectively, and the tournament selection size is 7 (these settings are recommended in the GP literature [14]). The evolution is allowed to run for maximum of 50 generations or terminated early if a solutions with optimal fitness is found. In practice, the second constraint was never satisfied on the training set. We also experimented with several settings for the maximum number of generations (25, 50 and 100), but found very little (if any) improvement in fitness after 50 generations.

4   Experimental Results: GP for Answer Ranking

This section outlines the Watson data, experiment setup and evaluation measures, and presents the results using GP directly for answer ranking (to learn the feature subsets).

4.1 Experimental Data

A variant of the *Jeopardy!* challenge Watson pipeline [7] is used to generate the experiment data (candidate answers to be ranked). Two English question sets consisting of 1200 training and 698 test questions are used as input to the Watson system. Both sets contain general knowledge questions with factoid-type answers (such as the example shown in Sect. 3.1) in the ingested Watson corpus (in this case, English Wikipedia). This means that Watson searches over English Wikipedia for answers where source Wikipedia XML documents[2] are first ingested into the system (ingestion is the process of transforming and loading documents for use by Watson). These questions and answer sets were originally gathered by the Watson team [7]. In its ML phase, this Watson configuration uses a two-phase machine learning cascade, i.e., two connected binary logistic regression models, where each model is trained using the *full* feature set (354 features) to perform answer ranking [10].

The Watson scoring algorithms output 118 base features (floating point numbers) for each generated candidate answer. Each of these 118 base features are post-processed to further generate two derived features, giving a total feature list of 354 features [10]. The first derived feature denotes whether a given scoring algorithm fired or not (derived feature values are 1.0 or 0.0, respectively). The second derived feature is the normalized value of each base feature across all feature vectors (using the minimum and maximum feature values). Due to IBM disclosure purposes we are unable to provide details on the type of features used in Watson.

As mentioned (in Sect. 2.2), there is a large class imbalance in the data as many more incorrect candidate answers generated compared the relatively few correct answers for a question. The 1200 training questions generated 204829 total candidate answers, split into 2282 and 202547 correct and incorrect answers, respectively. The 698 test questions generated 164069 total candidate answers, split into 2023 and 162046 correct and incorrect answers, respectively.

4.2 Experiment Setup

The goal of these experiments is to evolve GP ranking functions of varying complexity/size and evaluate the accuracy of these evolved ranking functions when applied directly to the test questions. The experiments use different GP tree depth limits to control the complexity/size of the evolved functions, where the pool of evolved GP trees forms the basis for feature mining in the next section. To evaluate accuracy, we use the *correct@1* metric. This is the percentage of questions correctly answered by an evolved GP function over all questions in the test set, where a question is correctly answered if the top-ranked candidate answer has the *correct* class label. For IBM disclosure purposes we are unable to provide direct accuracy values from our experimental results. Instead, we provide GP's accuracy proportional to Watson's accuracy (both on the test set of 698 questions), as shown by Eq. (2).

---

[2] Obtained from: http://dumps.wikimedia.org/.

$$correct@1_{prop} = \frac{correct@1_{GP}}{correct@1_{Watson}} \times 100\% \tag{2}$$

In Eq. (2), the numerator is the accuracy of an evolved GP function, and the denominator is the accuracy of the Watson system used to generate the experiment data (from Sect. 4.1). The same training set (1200 questions) is used for Watson and GP. Values for Eq. (2) will be between 0 and 100 where the higher the value, the better the GP accuracy proportional to Watson.

4.3 GP Experiment Results

Table 1 shows the results of four GP configurations (GP tree depth limits of 4, 5, 8 and 9) when GP used for candidate answer ranking over 50 runs on **test set**. This includes the *correct@1* question accuracy, the GP tree node counts (indicating model size), and feature node counts in the evolved trees (each feature node is counted exactly once in a tree). Table 1 shows both the average and best-of-run results over 50 runs, where the best-of-run GP tree is selected based on highest train accuracy over the 50 runs. The evolved GP tree with the maximum achieved test accuracy over 50 runs shown in parenthesis (next to best-of-run results); this is included to contrast the selected best-of-run model with the best possible GP performance on the test set. We experimented with six GP tree depth limits (4–9) but omit 6 and 7 for brevity and because our main goal here is to highlight the lower and upper bounds for GP accuracy (using the fewest number of model features).

Table 1 shows that GP achieves very good accuracy in relation to Watson, in particular, the best-of-run evolved GP trees. These GP functions use *much fewer* features than Watson and only employ one function for ranking compared to multiple (logistic regression) models in Watson. The best-of-run evolved GP trees for depths 4 and 8 achieve 85.6 % and 89.9 % accuracy, respectively, using only 8 and 21 features, respectively, compared to Watson (354 features). Even though GP cannot achieve 100 % accuracy (to match Watson accuracy), these results nevertheless represent a significant feature reduction. These evolved GP function provide a valuable indication of the key features in Watson, which is the main goal of our investigation. For example, the 21 features identified in the

Table 1. Results of four GP configurations when GP used for candidate answer ranking over 50 runs on **test set** (± is standard deviation). The best-of-run GP tree over 50 runs is selected based on highest train accuracy (maximum achieved test accuracy over 50 runs shown in parenthesis).

GP Config.	Average over 50 runs			Best-of-run GP tree over 50 runs		
	Test Acc.	#Tree Nodes	#Feat. Nodes	Test Acc.	#Tree Nodes	#Feat. Nodes
Depth 4	82.3 ± 2.7	17.1 ± 3.3	6.8 ± 1.2	85.6 (90.0)	17 (15)	8 (7)
Depth 5	83.1 ± 2.3	29.7 ± 5.2	8.2 ± 1.5	82.2 (91.1)	24 (32)	10 (10)
Depth 8	83.7 ± 2.3	111.7 ± 39.5	26.6 ± 7.8	89.9 (91.7)	96 (132)	21 (31)
Depth 9	85.4 ± X.X	163.6 ± 85.2	32.2 ± 13.1	88.2 (93.7)	112 (128)	19 (25)

best-of-run GP function for depth 8 are responsible for roughly 90 % of Watson's question ranking accuracy and represent key components in the larger complex Watson system.

Regarding the GP configurations, as expected, a clear relationship can be seen between the tree depth limit and accuracy in Table 1. Here larger evolved GP trees (typically) have better accuracies (since larger trees contain more useful features). There is very little difference in the average GP accuracy when maximum tree depth is 4, 5 and 8. This suggests a performance plateau for tree depths $<= 8$, where larger more complex GP trees do not perform much better than smaller trees (on average). Table 1 also shows the the average number of total nodes and feature nodes in the evolved GP trees are both much lower than the maximum expected values (estimated in Sect. 3.3).

5   Experimental Results: Feature Subsets in Watson

This section outlines the approaches to automatically build the feature subsets and the results of the Watson system using these generated feature subsets.

5.1   Constructing the Feature Sets

The previous section evolved GP functions of varying complexity/sizes and applied these functions directly to the rank the test questions. This section does feature mining of these evolved GP trees to automatically extract feature subsets to use directly in the Watson system. In other words, Watson will use is own machine learning framework to rank the test questions (as outlined in Sect. 2.1) but using a limited feature set (and not the full 354 Watson features). The goal of these experiments is to evaluate the performance of a learned feature subset in the Watson system. Two approaches are compared to automatically build feature subsets from the evolved GP trees. Each approach constructs two differently sized feature sets (a small feature set using at most 10 features, and a larger set using closer to 20 features) to compare different levels of feature reduction on question accuracy in the Watson system.

The first approach automatically extracts all features from the best-of-run evolved GP trees for a given GP configuration (from Sect. 4.3). Here one evolved GP tree produces one feature set, where the feature set size depends on the number of feature nodes in the GP tree. For example, the extracted feature set for the GP expression (+ (if f2 -2.0 f5) f4) will be {f2, f4, f5}. In these experiments, the two best-of-run evolved GP trees using maximum tree depths of 4 and 8 are selected (giving 8 and 21 features, respectively).

The second is a more consensus-based approach which takes the most frequently occurring N features from *all* evolved GP trees over all GP configurations. Here all feature nodes are counted across 300 evolved GP trees (50 runs × 6 GP configurations using different maximum tree depths). A feature node is counted exactly once in a tree (even if it occurs in multiple locations in a tree).

In these experiments, the two features sets are built using N values of 10 and 20, representing the top 10 and top 20 most commonly occurring features over all evolved GP trees.

For a comparison, we also evaluate two other approaches for feature selection. The first approach uses the Gram-Schmidt orthonormalising (GSO) numerical analysis procedure to rank the feature vectors by decreasing relevance to output (class label), where the top N features represent the N most relevant features. The second approach uses the WEKA toolkit [11], which provides a broad array of algorithms for this purpose. Here attribute selection consists of pre-processing the data set, selecting an algorithm for evaluating individual attributes and its corresponding parameters, selecting a method for performing the search using the evaluator, and deciding whether to perform cross-validation or to run on the whole dataset. During pre-processing, the data was re-sampled to balance the classes, and to reduce the size to roughly 15000 instances. The `GreedyStepwise` attribute selection method was applied in the forward direction (to incrementally add the attribute that most improves the prediction), stopping when additional attributes fail to yield additional gains. This algorithm produces a ranked list of attributes. Finally, the `ClassifierSubsetEval` attribute evaluator algorithm paired with the `Logistic` classifier trainer was run, limiting the number of iterations to 10 (with 5-fold cross-validation in WEKA).

5.2 Experiment Results

Table 2 shows the experimental results using eight different feature subsets directly in the Watson system on the test set (698 questions). These feature subsets are categorized into small sets (exactly 10 features) and larger sets (closer to 20 features) and are constructed using features derived from GP, WEKA and GSO. The features extracted from the two GP strategies, that is, from the best-of-run GP trees and and all evolved GP trees, are denoted by *GP-Best* and *GP-All*, respectively, in Table 2. In each experiment, the Watson system was retrained using the given feature subset only. As the logistic regression model training process in Watson is deterministic, only one training run is required. Similar to the previous results, the accuracy values in Table 2 are proportional to the full Watson system using all features.

Table 2. Question accuracy (on the test set) of the Watson system using feature subsets.

Small feature sets (approx. 10 features)			Larger feature set (approx. 20 features)		
Feature Selection	# Features	Accuracy %	Feature Selection	# Features	Accuracy %
GP-Best (Dep. 4)	8	**91.9**	GP-Best (Dep. 8)	21	88.7
GP-All (Top 10)	10	91.6	GP-All (Top 20)	20	93.7
WEKA Top 10	10	75.0	WEKA Top 20	18	87.9
GSO Top 10	10	89.5	GSO Top 20	20	**94.2**

In Table 2, the best accuracy for the small and larger feature subsets are achieved by *GP-Best (Dep 4)* and *GSO Top 20*, that is, 91.9 % and 94.2 %, respectively (highlighted in bold). Notice that feature sets based on the GP-Best approach achieved similarly good accuracies across both the smaller and larger feature set groups (these are only slightly lower than the best accuracies in Table 2). In contrast, the WEKA-based feature subsets show the poorest performance for both feature set groups; while GSO and GP-Best both perform well on one feature set group but not the other. This makes sense intuitively as the most frequently occurring feature combinations are based on a consensus over many different evolved GP ranking functions. This also suggests that extracting the most frequently occurring features across all evolved GP trees is an effective strategy to find small and accurate feature subsets compared to Weka and GSO.

It is interesting that the best-performing GP functions when applied directly for ranking as a "stand-alone" model (from Table 1) achieves competitive performances on the same test questions relative to the Watson system limited to the smaller feature subsets (from Table 2) but still using a cascade of logistic regression models. For example, the best-evolved GP tree of depth 8 achieves 89.9 % accuracy as a "stand-alone" ranking model, but the Watson system limited to the same 21 features from this tree has 88.7 % accuracy. This is likely due to the expressive power of the evolved GP ranking functions, in particular, the feature construction aspect (arithmetic operators in the GP trees can manipulate feature values). This is not available to the logistic regression classifiers, even when applied in successive phases in the cascade.

However, no evolved GP tree is able to achieve the best accuracy in Table 2, that is, 94.2 % from Watson using the GSO Top 20 feature subset (the maximum test accuracy from Table 1 is 93.7 for GP configuration using depth 9). This might be due to several factors such as the GP configuration (maximum tree depth too restrictive), and only single "stand-alone" ranking model (not a cascade of models). A full comparison between Watson and GP where GP is not limited in program complexity and/or is applied in successive phases is outside the scope of this paper and will be future work.

6   Conclusions and Future Work

The main goal of this paper is to develop a two-phase GP approach to find the most useful features in IBM Watson's NLP pipeline. GP is first used to evolve simple but accurate functions to classify and rank candidate answers generated by Watson in response to question. The evolved GP functions are then mined to automatically extract the most frequently occurring features as these represent features automatically identified from the learning process. This goal was achieved by examining the question accuracy of the evolved GP functions on the question ranking data, automatically extracting several feature subsets from the evolved GP trees, and evaluating the performance of these feature subsets directly in the Watson system.

Experiments show that the best-evolved GP functions achieve accuracies to within 93 % of the Watson system but used *much fewer* features (fewer than 8 % of Watson features). Some evolved GP functions also outperformed the Watson system limited to the same feature subsets (from the GP functions) and using multiple logistic regression models. This demonstrates the expressive power of the relatively simply evolved GP ranking models compared to the logistic regression cascade. Experiments also show that a consensus-based approach for extracting the most frequently occurring N features from the evolved GP trees can effectively find features that perform well consistently across two feature group sizes; whereas other approaches to build the feature subsets tend to perform well on one feature group size but not other. As Watson is an open-domain system, the key features identified by our GP approach is relative to the factoid questions we asked of Watson. However, our approach is a generalised methodology to find useful features that can easily be applied to any question/customer domain.

For future work our next steps involve a deeper analysis of these learned feature subsets. This includes a careful analysis of the evolved GP trees to find commonly occurring *composite* features in the trees. We will investigate a parsimony objective in the fitness function to explicitly favour the evolution of small GP trees, and also explore different training configurations in GP and Watson to improve question accuracy using smaller feature subsets.

Acknowledgements. We would like to thank IBM Research Staff members Dr. Vittorio Castelli and Dr J. William Murdock for their valuable contributions to this paper.

References

1. Bhowan, U., Johnston, M., Zhang, M., Yao, X.: Reusing genetic programming for ensemble selection in classification of unbalanced data. IEEE Trans. Evol. Comput. **18**(6), 893–908 (2014)
2. Bhowan, U., Johnston, M., Zhang, M., Yao, X.: Evolving diverse ensembles using genetic programming for classification with unbalanced data. IEEE Trans. Evol. Comput. **17**(3), 368–386 (2012)
3. Davis, R.A., Charlton, A.J., Oehlschlager, S., Wilson, J.C.: Novel feature selection method for genetic programming using metabolomic 1 H NMR data. Chemom. Intell. Lab. Syst. **81**(1), 50–59 (2006)
4. Espejo, P., Ventura, S., Herrera, F.: A survey on the application of genetic programming to classification. IEEE Trans. Syst. Man Cybern. Part C Appl. Rev. **40**(2), 121–144 (2010)
5. Fan, W., Gordon, M.D., Pathak, P.: Discovery of context-specific ranking functions for effective information retrieval using genetic programming. IEEE Trans. Knowl. Data Eng. **16**(4), 523–527 (2004)
6. Fan, W., Gordon, M.D., Pathak, P.: A generic ranking function discovery framework by genetic programming for information retrieval. Inf. Process. Manage. **40**(4), 587–602 (2004)
7. Ferrucci, D., Brown, E., Chu-Carroll, J., Fan, J., Gondek, D., Kalyanpur, A.A., Lally, A., Murdock, J.W., Nyberg, E., Prager, J., et al.: Building Watson: an overview of the DeepQA project. AI Mag. **31**(3), 59–79 (2010)

8. Ferrucci, D., Levas, A., Bagchi, S., Gondek, D., Mueller, E.T.: Watson: beyond Jeopardy!. Artif. Intell. **199**, 93–105 (2013)
9. Ferrucci, D.A.: Introduction to "This is Watson". IBM J. Res. Dev. **56**(3.4), 1:1–1:15 (2012)
10. Gondek, D., Lally, A., Kalyanpur, A., Murdock, J.W., Duboue, P.A., Zhang, L., Pan, Y., Qiu, Z., Welty, C.: A framework for merging and ranking of answers in DeepQA. IBM J. Res. Dev. **56**(3.4), 14:1–14:12 (2012)
11. Hall, M., Frank, E., Holmes, G., Pfahringer, B., Reutemann, P., Witten, I.H.: The WEKA data mining software: an update. In: SIGKDD Explorations. vol. 11 (2009)
12. Koza, J.R.: Genetic Programming: on the programming of computers by means of natural selection, vol. 1. MIT Press, Cambridge (1992)
13. Muharram, M., Smith, G.: Evolutionary constructive induction. IEEE Trans. Knowl. Data Eng. **17**(11), 1518–1528 (2005)
14. Poli, R., Langdon, W.B., McPhee, N.F., Koza, J.R.: A field guide to genetic programming (2008). Lulu.com
15. Tan, X., Bhanu, B., Lin, Y.: Fingerprint classification based on learned features. IEEE Trans. Syst. Man Cybern. Part C Appl. Rev. **35**(3), 287–300 (2005)
16. Trotman, A.: Learning to rank. Inf. Retrieval **8**(3), 359–381 (2005)
17. Wang, L., Fan, W., Yang, R., Xi, W., Luo, M., Zhou, Y., Fox, E.A.: Ranking function discovery by genetic programming for robust retrieval. In: TREC. pp. 828–836 (2003)
18. Yeh, J.Y., Lin, J.Y., Ke, H.R., Yang, W.P.: Learning to rank for information retrieval using genetic programming. In: SIGIR Workshop: Learning to Rank for Information Retrieval (2007)

Automatic Evolution of Parallel Recursive Programs

Gopinath Chennupati[(✉)], R. Muhammad Atif Azad, and Conor Ryan

Bio-Computing and Developmental Systems Group, Computer Science
and Information Systems Department, University of Limerick,
Limerick, Ireland
{gopinath.chennupati,atif.azad,conor.ryan}@ul.ie

Abstract. Writing recursive programs for fine-grained task-level execution on parallel architectures, such as the current generation of multi-core machines, often require the application of skilled parallelization knowledge to fully realize the potential of the hardware. This paper automates the process by using Grammatical Evolution (GE) to exploit the multi-cores through the evolution of *natively* parallel programs. We present Multi-core Grammatical Evolution (MCGE-II), which employs GE and OpenMP specific pragmatic information to automatically evolve *task-level* parallel recursive programs. MCGE-II is evaluated on six recursive C programs, and we show that it solves each of them using parallel code. We further show that MCGE-II significantly decreases the parallel computational effort as the number of cores increase, when tested on an Intel processor.

Keywords: Grammatical evolution · Automatic parallelization · Recursion · Program synthesis · OpenMP · Evolutionary auto-parallelization

1 Introduction

The advent of multi-core (2 or more) processors for PCs has been little short of a revolution in terms of price/performance ratio. Multi-core processors are integrated with multiple processing elements on a single chip.

However, the actual improvement experienced often depends on the way that the parallel programs are coded. With a small number of cores, single processes or Virtual Machines can occupy each, but, as multi-core becomes *many-core*[1], the operating systems face difficulty in utilizing the power of all the cores.

High performance computing researchers manifested this so-called multi-core menace as the *third software crisis* [3], the imminent inability to program and fully exploit multi-cores. In accordance, Gartner [4] also predicted that software is trailing the surge of multi-cores, and urged the need for the development of computer applications that can ease the difficulty in programming them.

---

[1] For example, the Intel Polaris has 80 cores, while the picoChip PC200 has 200+.

© Springer International Publishing Switzerland 2015
P. Machado et al. (Eds.): EuroGP 2015, LNCS 9025, pp. 167–178, 2015.
DOI: 10.1007/978-3-319-16501-1_14

Achieving parallelism is hard considering its challenges, uppermost of which is that, in general, programmers are trained to think and write sequential programs. Other significant roadblocks include the difficulties involved in code and data restructuring, race and deadlock occurrences, debugging parallel programs and attaining behavioural equivalence of both serial and parallel programs. Evolutionary Computation (EC), on the other hand, has less baggage than its human counterparts, making parallel programs an ideal target. In this paper, we propose to apply EC to evolve parallel programs that optimize a non-functional property (time) while we also produce a qualitative solution to a given problem.

We employ Grammatical Evolution (GE) [15] to evolve C programs that use OpenMP API [12]; this results in the evolution of a complete parallel program. Our approach obviates the need for programmers to think in a parallel manner while still letting them produce parallel code using essentially the same techniques as are used in standard GE and GP. Also, to the best of our knowledge it is the first evolutionary attempt to evolve a natively parallel recursive program.

The rest of the paper is outlined as follows: Sect. 2 introduces a motivating example, discusses the literature on evolutionary attempts on recursion and *automatic parallel code generation*; Sect. 3 details our approach in automatic evolution of parallel programs; Sect. 4 presents the experimental methodology and results; and finally, Sect. 5 concludes and outlines the future aspirations.

2   Background

In this section, we explore the scope for parallel recursion, evolution of recursive programs and, evolutionary parallel code generation.

2.1   Scope for Parallel Recursion

Recursion is a method of making self referential calls, widely used to solve a problem by breaking it into smaller sub-problems, a divide-and-conquer strategy.

```
int fib(int n){
    if(n <= 2)    return n;
    else          return fib(n−1) + fib(n−2);
}
```

Fig. 1. Motivating example that generates *Fibonacci* sequence recursively.

Consider a simple recursive program that generates a Fibonacci sequence, as in [9], shown in Fig. 1. The procedure *fib* terminates upon fulfilling the *base case*. The two *independent* recursive calls follow with input decreasing by 1 and 2 in the first and second call respectively. These recursive calls can be computed simultaneously, thus allowing to execute the two calls in parallel. The parallelism exploited can be fine-grained, where both the calls are computed before the final addition happens. Also, note that since the two recursive calls generate different execution traces, their concurrent execution represents *task level* parallelism.

The challenge then is to *automatically* discover such a scope for parallel recursion through Machine Learning techniques; we explore Grammatical Evolution (GE) to this end in this paper. However, first we review literature on evolving recursive programs regardless of parallelism in Sect. 2.2 and then, in Sect. 2.3 we review approaches to evolve parallel programs.

2.2   Evolution of Recursive Programs

EC based attempts on automatic evolution of recursive programs were initiated by Koza [9] with the introduction of a sequence reference function, *(SRF K D)* for Fibonacci sequence. The function referenced the previously evaluated values from a table and returned the Kth value if available, otherwise, default value D.

Brave [6] explored tree search on a simple planning problem through a restrictive form of recursion using Automatically Defined Functions (ADFs). To prevent excessive recursive calls, the ADFs used were only allowed as many recursive calls as the tree depth. Whigham and McKay [19] used tree based GP to learn recursive functions that take as input an element of a list and its position. The function returned true if the element was found in the list, and NIL (false) otherwise. However, they concluded that evolving recursion was inappropriate for automated learning because of infinite calls.

In an attempt to discourage infinite recursion, Wong and Mun [21] used an adaptive grammar based GP by adjusting the *weights* associated with the production rules of the grammar. This approach increased the probability of success and decreased the number of infinite-recursive programs. Yu and Clark [22] used *implicit recursion* in performance gains in GP. Implicit recursion materialized through a higher order function that took two arguments, a binary operator and a list of values; the operator then is placed in between successive pairs of items of the list and evaluated from left to right.

Recent EC literature also shows renewed efforts to automatic recursion. Among them, Spector et al., [17] evolved recursive programs using PushGP. PushGP supports explicit manipulation of iterative and recursive routines. Agapitos and Lucas [1,2] analysed the generality of evolving modular recursive sorting algorithms with the help of Object Oriented Genetic Programming (OOGP) by defining special classes and methods in Java. Moraglio et al., [10] presented a general non-recursive scaffolding method that evolved a recursive list reversal program using a context free grammar based GP.

2.3   Evolutionary Generation of Parallel Code

The generation of parallel code can, broadly speaking, be divided into two categories: *auto-parallelization of sequential code* and the *generation of native parallel code*. Auto-parallelization mirrors the approach taken by many programmers when they generate parallel code. That is, they first identify the parallelism in an algorithm through an analysis. Examples such as *Automatic Polyhedral Parallelizer* [5] converts C programs to multi-core executable OpenMP programs, many conventional approaches operate in this way.

Evolutionary auto-parallelization of serial code was initiated by Walsh and Ryan using GP, *PARAGEN-I* [[13], Chapter-5] automatically mapped the serial programs onto parallel hardware; however, handling complex loops remained a major challenge. Then, they extended it as *PARAGEN-II* [[13], Chapter-7] to evolve transformations that first analysed serial-dependency of instructions; this analysis was performed during the fitness evaluation. A further attempt [16] extended *PARAGEN-II* to perform *loop fusion*, that is, merge independent tasks of different loops into a single loop. All the experiments in [13,16] were carried out on a Beowulf cluster using a distributed memory model.

Similar approaches for GAs include: Nisbet [11], presented the Genetic Algorithm Parallelization System (GAPS) that dealt with the optimization of transformations while, Williams [20] proposed *REVOLVER* with two representations (*gene-transformation, gene-statement*) to transform loops and programs.

The preceding approaches rely on the existence of a **working program** that they modify. In contrast, *natively parallel code generation* solves two problems together: generate a working program which is also parallel. For example Trenaman [18] showed automatic design of controllers for autonomous agents using a multi-tree GP representation that evolved concurrent execution of the agents.

With the advent of modern *multi/many-core* architectures, and the so called *death of scaling*[2] parallel code generation is critical to performance scaling. Realizing that Chennupati et al., [7] innovated with *Multi-Core Grammatical Evolution* (MCGE) by evolving multi-core based parallel programs for two well known GP regression problems. They also analysed execution times of the evolved programs in [8]. The next section elaborates on the opportunities, both in hardware and software for EC to advance multi-core computing and presents the new approach that builds on MCGE (or MCGE-I as we term it here).

3   Multi-core Grammatical Evolution (MCGE)

Multi-core processors are now commonplace, where the operating system treats each core as an independent execution entity. All multi-core processors *share memory* to interact and synchronize among themselves.

The general approach to exploit parallelism in shared memory models is to write *fork/join* programs. In that a *master* thread spawns *slave* threads on all cores and then joins them back after the completion of their task. In this paper, we use the OpenMP that uses the fork/join model. Next, we describe OpenMP.

3.1   OpenMP

OpenMP [12] is a portable, scalable directive based specification to write parallel programs on shared memory systems, jointly defined by major computer hardware and software vendors. It has compiler directives, environment variables, and run time libraries that combine to parallelize code in C/C++ and Fortran. OpenMP programs allow both shared and thread-private data structures.

---

[2] http://www.gotw.ca/publications/concurrency-ddj.htm.

The general syntax of an OpenMP directive can be seen as follows:

$\#pragma\ omp\ directive\_name\ [clause[[,]clause]\dots]\ \{< statement\ block >\}$

where, a *directive_name* can be replaced with any one of the constructs: *parallel for*, *parallel sections*, *master* and synchronization (*critical*). The number of threads allowed in a given parallel region depends on the clauses that a directive allows to control. A complete description of OpenMP can be found in [12].

$\langle omppragma \rangle$	::= #pragma omp parallel for $\langle newline \rangle$ '{' $\langle parcode \rangle$ | #pragma omp parallel $\langle newline \rangle$ '{' $\langle parcode \rangle$ | #pragma omp parallel sections $\langle newline \rangle$ '{' $\langle parblocks \rangle$
$\langle parcode \rangle$	::= if($\langle var \rangle$ $\langle lop \rangle$ $\langle const \rangle$) '{'int a = $\langle expr \rangle$; res $\langle bop \rangle$= a; $\langle newline \rangle$'}' else '{' int a = $\langle expr \rangle$; $\langle newline \rangle$ res $\langle bop \rangle$= a; '}' $\langle newline \rangle$ '}' $\langle newline \rangle$ $\langle result \rangle$ $\langle newline \rangle$ '}'
$\langle parblocks \rangle$	::= if($\langle var \rangle$ $\langle lop \rangle$ $\langle const \rangle$) '{'$\langle newline \rangle$ int a = $\langle expr \rangle$; '}' else '{' $\langle newline \rangle$ $\langle blocks \rangle$ $\langle newline \rangle$ '}'
$\langle blocks \rangle$	::= $\langle blocks \rangle$ | $\langle blocks \rangle$ $\langle newline \rangle$ $\langle blocks \rangle$ | #pragma omp section $\langle newline \rangle$ '{' int a = $\langle stmt \rangle$; $\langle newline \rangle$ #pragma omp atomic $\langle newline \rangle$ res $\langle bop \rangle$= a; '}' $\langle newline \rangle$ '}' $\langle newline \rangle$ $\langle result \rangle$ $\langle newline \rangle$
$\langle result \rangle$	::= return $\langle var \rangle$;
$\langle expr \rangle$	::= $\langle var \rangle$ | $\langle stmt \rangle$ | $\langle stmt \rangle$ $\langle bop \rangle$ $\langle stmt \rangle$
$\langle stmt \rangle$	::= fib($\langle var \rangle$ $\langle bop \rangle$ $\langle const \rangle$)
$\langle bop \rangle$	::= + | - | * | /
$\langle lop \rangle$	::= '>=' | '<=' | '>' | '<' | '=='
$\langle const \rangle$	::= 0 | 1 | 2 | 3 | 4 | 5 | 6 | 7 | 8 | 9
$\langle var \rangle$	::= n | res
$\langle newline \rangle$	::= \n

Fig. 2. MCGE-II grammar that generates a recursive Fibonacci sequence program, where OpenMP parallelization pragmatic information is included.

3.2   MCGE-II

Unlike MCGE-I that evolved programs exhibiting *data parallelism* (identical sequence of instructions operating on different data), MCGE-II evolves parallel programs that exhibit fine-grained *task level parallelism* (different sub-tasks executing in parallel).

Note, MCGE-II does not make any changes to the design of GE. Instead, it relies on grammars so designed as to embody the knowledge that allows GE to produce parallel recursive programs. The grammars allow GE to select from various OpenMP pragma directives. Unlike PARAGEN-II [13], MCGE-II does not utilise dependency analysis; instead, GE works the data interdependencies out by selecting pragmas that suite program correctness. Figure 2 presents one such grammar for the Fibonacci sequence problem described in Sect. 2.1.

In Fig. 2 the non-terminal symbol $< stmt >$ maps to problem specific recursive call (fib in this case). Any choice of the non-terminal $< omppragma >$ ensures that any recursive calls they *enblock* execute in parallel, but, the last choice (i.e., parallel sections) particularly suites task level parallelism. However, it is down to evolution to select this directive preferentially. The syntax of this pragma requires the use of a special pragma ($\#pragma\ omp\ section$) that designates blocks of code to execute as separate tasks.

Although we evolve programs in C, MCGE-II is general enough to apply to other programming languages that offer OpenMP like parallelization.

4 Experiments

We test the proposed approach on six well known recursive problems as summarized in Table 1. The solutions to the problems require several interesting features of a parallel recursive solution such as branching, recursive calls, recursive calls over arrays, and temporary, shared and/or private (OpenMP specific) variables. The results also indicate that the difficulty of the problems varies.

In all the experiments, we maximise fitness by first computing the mean absolute difference between the desired and evolved output and then normalizing it between 0 and 1. For the problems that take an array as input, we generate 100 random integers that are in the range 0 to 100. For the single input problems, we randomly select a value from the range 1 to 50; it is large enough to expand the

Table 1. The problem sets used in the experiments and their properties.

#	Problem	Description	Input	Local Variables
1	Sum-of-N	Sum of first N numbers	int	1
2	Factorial	Factorial of a given number	int	1
3	Fibonacci	Generate a fibonacci sequence	int	2
4	Binary-Sum	Add pairs of elements in an array	int [ ], int, int	2
5	Reverse	Reverse the array/list of elements	int [ ], int, int	2
6	Quicksort	Sorts an array in ascending order	int [ ], int, int	3

execution trace. For example, to compute the Fibonacci sequence for 50 we have to get the sequence right for the smaller inputs as well. We take the non-evolutionary result of the respective problem as output of that problem.

The experimental results that we report in this paper are carried out using the default experimental settings of GE, with a population size of 500 individuals for all the problems. We use *Sensible Initialisation* [14], a ramped half and half approach to initialize derivation trees in GE, where the minimum depth of the derivation tree is 9, and the maximum depth is 15. We use one-point crossover with a probability of 0.9, and a per bit mutation with a probability of 0.01. We use a steady state GA where the best individuals replace the worst in the population. The results reported in this paper base on a total of 50 runs per setting, with each run lasting for 100 generations. The experiments are conducted on an Intel (R) Xeon (R) CPU E7-4820 (16 cores) while the evolved programs are evaluated using GNU GCC compiler with *-fopenmp* option.

4.1 Terminating Recursion

Preventing infinite recursion is crucial in automatic generation of recursive programs. To this end, we limit the maximum number of recursive calls to the maximum number of generations allowed to GE. If a program exhausts this quota, the evolved function simply returns the input value and terminates; otherwise, the function returns the computed value. Although, the actual limit is an *ad-hoc* choice, it is akin to setting a maximum *size limit* in standard GP.

Furthermore, we investigate three different approaches in formulating the termination (for base case). The first approach, referred to as *const-10* henceforth, allows the condition as well as the recursive calls to entail a constant with any one of the 10 constants ($< const >$) ranging between 0 and 9 (both inclusive).

The second approach, termed as *cond* evolves the terminating condition so that it always compares with 1, that is $< condition >::= if (< var >< lop > 1)$. Thus, only one constant is available to choose from, however, the recursive calls are allowed to choose from all the 10 available constants in $< const >$.

Finally, the third approach, termed as *const-limit*, where the constants ($< const >$) range is reduced from $(0 \ldots 9)$ to $(1, 2)$. That is, both the base case and the recursive calls can only choose one of the two available constants, 1 and 2. It is evident that we reduce the search space. However, the last two approaches incorporate problem specific knowledge into the grammar to facilitate evolution.

4.2 Experimental Results

We report two key statistics in this section: the mean best fitness and the mean of the total execution time for all the best of generation programs.

Mean Best Fitness: Figure 3 presents and compares the mean best fitness for the 6 experimental problems; the results are averaged across 50 runs at each generation for all the three variations. On all the six problems, a Student's *t-test* at $\alpha = 0.05$ shows insignificant difference between *const-10* and *cond*. However, the results for *cond* improve slightly late in the runs.

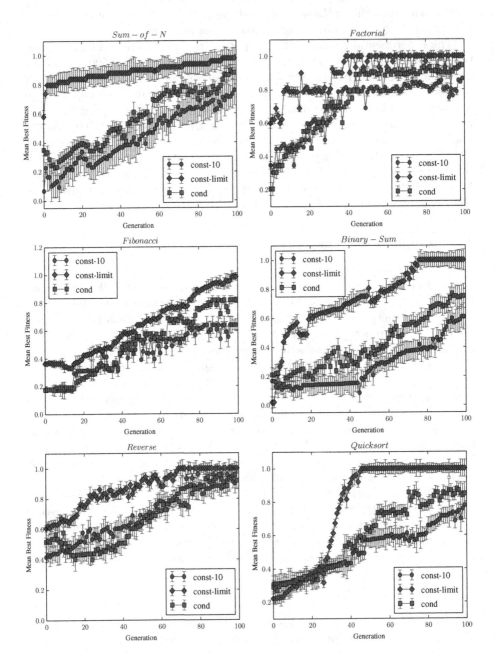

Fig. 3. Mean best fitness graphs (with standard deviation) of MCGE-II for all the six experimental problems that are averaged across 50 runs of 100 generations.

Mean best fitness of *const-limit* proved to be *significantly* different from the rest as measured with the Student's *t-test* at $\alpha = 0.05$. In fact, *const-limit* solved all the problems. An obvious reason for that is that the constants used for generating the base case are limited to suite the problem.

The results indicate that except for *Sum-of-N*, the problems tackled here are not easy: even with *const-limit* it takes at least 40 generations to find the correct solution. We observed that although, syntactically an almost-ideal solution can be generated quite early, finding the exact solution can take a long time.

For example, in the case of *const-limit*, the best evolved Fibonacci program in generation 17 is given as follows, that has a fitness value of 0.36,

```
#pragma omp parallel {
    if (n > 2) { int a = n; res += a; }
    else    { int a = fib (n-2) * fib (n-1);
    res += a; }
} return res;
```

and, at generation 98, we find the following individual that has fitness 1.

```
#pragma omp parallel sections {
    if (n <= 2)   { #pragma omp section { int a = n;
            #pragma omp atomic                    res += a;
    } }
    else   {   #pragma omp section { int a = fib (n-1);
                        #pragma omp atomic   res += a; }
                #pragma omp section { int a = fib (n-2);
                        #pragma omp atomic   res += a; }    }
} return res;
```

In the ideal program, the logical operator $>$ changed to $<=$ in the *if* condition and, the binary operator $*$ to $+$, and now sums the result of recursive calls. Moreover, the parallelism exerting pragmas also changed: the pragma (*parallel*) in the first program creates multiple threads and executes the program code that it enblocks in parallel. Although, the evolved program (with its binary operation ($*$)) is not an optimal solution, it is even worse with the chosen pragma, because it computes the same (and wrong) process twice in two threads. The ideal program, uses a correct pragma (*parallel sections*), and also identifies tasks to execute separately in threads; these tasks are enblocked with the pragma *sections*. Notice, the shared variable *res* is updated atomically, thus preventing race conditions. This perfectly achieves task level parallelism, thus instantiating a parallel recursive program.

We can try another approach that first finds a correct solution and then parallelize it. This can be encouraged by incorporating even more problem specific information. In the absence of such knowledge we can use a multi-staged approach where the first stage finds the correct solution while the second stage maximizes the parallelism; however, we leave that to future work.

Execution Time: The advantage of multi-cores is to scale up the performance through parallel processing. Therefore we report the execution time taken by the

Table 2. Mean best execution time (in secs) (mean [standard deviation]) of MCGE-II, averaged across 100 generations of 50 runs. We record the execution time for a varying number of cores on an Intel Xeon multi-core processor.

Problem	Cores				
	1	2	4	8	16
Sum-of-N	1884.31[15.46]	1782.16[16.01]	551.02[18.42]	628.61[25.27]	331.89[36.24]
Factorial	2481.48[37.09]	2988.14[22.31]	781.21[26.32]	724.18[29.74]	487.76[35.61]
Fibonacci	3799.31[41.14]	2821.27[37.32]	1409.54[34.11]	921.39[21.92]	608.18[49.17]
Binary-Sum	3336.24[67.76]	2134.92[29.16]	1683.86[29.22]	738.99[37.32]	482.48[27.46]
Reverse	3222.69[64.76]	3348.59[38.41]	1035.77[43.36]	596.62[35.17]	520.11[63.19]
QuickSort	4644.87[29.04]	2578.19[27.11]	1540.07[63.69]	705.78[44.51]	577.18[41.28]

evolving individuals that in this section represents the *total* time required to execute *all* the best of generation programs produced in a run; we report the mean of this time (averaged over 50 runs). We term it *mean best execution time*.

Note, that the execution time of an evolved program depends on various factors such as the number of users using the resources, and the level of parallelism exerted. We do not delineate the effect of different factors in this study and report the total time and observe the impact of parallelism across an increasing number of cores. We use the OpenMP timer utility function *omp_get_wtime()* to record the execution time for each individual.

Table 2 presents the mean best execution time of the MCGE-II evolved programs for the six experimental problems that are averaged across 50 runs with each run containing 100 generations. We also record the execution time as we increase the number of cores from 2 to 16.

The results show that the time reduces significantly as we increase the number of cores; again, we measure the significance with the t-test at $\alpha = 0.05$, except when we increase the cores from 1 to 2. [8] discusses why increasing from 1 to 2 cores fails to yield the expected gains; the reasons include scheduling overhead nullifying the gain in speed-up through two cores, as well as somewhat longer genomes of the OpenMP enabled solutions that use pragma directives (unlike the normal individuals which do not have this additional code).

5   Conclusions

In summary, this study clearly showed the evolution of parallel recursive programs that employed the OpenMP directives along with the GE context free grammars. The presented results explored various possibilities of evolving parallel programs while also maintained solution correctness; this was a dual challenge, a challenge that an EC approach tackled successfully given the problem suite. We noticed that the attempts to prevent infinite recursion, although ad-hoc, did not prevent evolution from exploring the high quality solution space.

We also presented the computational speed up exploited by the evolved programs on multi-cores. This is a first attempt to automatically solve the problem so significant as to be described as the *third software crisis*.

Future work can expand in several directions: we intend to see the impact of parallel recursion on code growth so as to delineate the computational overhead of scheduling threads (similar to [8]). Furthermore, we aim to extend beyond C and explore the potential of interpreted languages for evolving parallel code. This will decrease the time taken to complete the evolutionary simulations that currently rely on external system calls to write the evolving programs on to the disk, compile and then execute it as an external process. Finally, we can use evolution to *promote* parallelism: in this study we did not prefer individuals that exploited maximum parallelism. Challenges associate such an undertaking are avoiding *excessive* parallelism which can result from producing too many threads that accomplish too little individually, and maintaining correctness of the evolving solution while promoting parallelism.

References

1. Agapitos, A., Lucas, S.M.: Evolving efficient recursive sorting algorithms. In: IEEE Congress on Evolutionary Computation, pp. 2677–2684. IEEE (2006)
2. Agapitos, A., Lucas, S.: Evolving modular recursive sorting algorithms. In: Ebner, M., O'Neill, M., Ekárt, A., Vanneschi, L., Esparcia-Alcázar, A.I. (eds.) EuroGP 2007. LNCS, vol. 4445, pp. 301–310. Springer, Heidelberg (2007)
3. Amarasinghe, S.: (How) can programmers conquer the multicore menace?. In: Proceedings of the International Conference on Parallel Architectures and Compilation Techniques, pp. 133–133. ACM (2008)
4. Barker, C.: Gartner: Multicore chips leave software trailing (Jan 2009)
5. Bondhugula, U., Hartono, A., Ramanujam, J., Sadayappan, P.: A practical automatic polyhedral parallelizer and locality optimizer. In: ACM SIGPLAN Conference on Programming Languages Design and Implementation. vol. 43, pp. 101–113. ACM (2008)
6. Brave, S.: Evolving recursive programs for tree search. In: Angeline, P.J., Kinnear Jr., K.E. (eds.) Advances in Genetic Programming, vol. 2, pp. 203–220. MIT Press, Cambridge (1996)
7. Chennupati, G., Azad, R.M.A., Ryan, C.: Multi-core GE: automatic evolution of CPU based multi-core parallel programs. In: Proceedings of the Genetic and Evolutionary Computation Conference Companion, pp. 1041–1044. ACM (2014)
8. Chennupati, G., Fitzgerald, J., Ryan, C.: On the efficiency of multi-core grammatical evolution (MCGE) evolving multi-core parallel programs. In: Proceedings of the Sixth World Congress on Nature and Biologically Inspired Computing (NaBIC), pp. 238–243. IEEE (2014)
9. Koza, J.R.: Genetic Programming: On the Programming of Computers by Means of Natural Selection. MIT Press, Cambridge (1992)
10. Moraglio, A., Otero, F.E.B., Johnson, C.G., Thompson, S., Freitas, A.A.: Evolving recursive programs using non-recursive scaffolding. In: IEEE Congress on Evolutionary Computation, pp. 1–8. IEEE (2012)
11. Nisbet, A.: A compiler framework for genetic algorithm (GA) optimised parallelisation. In: Bubak, M., Hertzberger, B., Sloot, P.M.A. (eds.) HPCN-Europe 1998. LNCS, vol. 1401, pp. 987–989. Springer, Heidelberg (1998)

12. OpenMP Architecture Review Board: OpenMP application program interface version 3.0., May 2008. http://www.openmp.org/mp-documents/spec30.pdf
13. Ryan, C.: Automatic Re-engineering of Software Using Genetic Programming, Genetic Programming, vol. 2. Springer, New York (1999)
14. Ryan, C., Azad, R.M.A.: Sensible initialisation in grammatical evolution. In: Barry, A.M. (ed.) Proceedings of the Bird of a Feather Workshops, Genetic and Evolutionary Computation Conference, pp. 142–145. AAAI (2003)
15. Ryan, C., Collins, J.J., Neill, M.O.: Grammatical evolution: evolving programs for an arbitrary language. In: Banzhaf, W., Poli, R., Schoenauer, M., Fogarty, T.C. (eds.) EuroGP 1998. LNCS, vol. 1391, pp. 83–95. Springer, Heidelberg (1998)
16. Ryan, C., Ivan, L.: Automatic parallelization of arbitrary programs. In: Langdon, W.B., Fogarty, T.C., Nordin, P., Poli, R. (eds.) EuroGP 1999. LNCS, vol. 1598, pp. 244–254. Springer, Heidelberg (1999)
17. Spector, L., Klein, J., Keijzer, M.: The push3 execution stack and the evolution of control. In: Proceedings of the Genetic and Evolutionary Computation Conference, pp. 1689–1696. ACM, New York (2005)
18. Trenaman, A.: Concurrent Genetic Programming, Tartarus and Dancing Agents. In: Langdon, W.B., Fogarty, T.C., Nordin, P., Poli, R. (eds.) EuroGP 1999. LNCS, vol. 1598, pp. 270–282. Springer, Heidelberg (1999)
19. Whigham, P.A., McKay, R.I.: Genetic approaches to learning recursive relations. In: Yao, X. (ed.) Progess in Evolutionary Computation. LNAI, vol. 956, pp. 17–27. Springer, Heidelberg (1995)
20. Williams, K.P.: Evolutionary algorithms for automatic parallelization. Ph.D. thesis, University of Reading (1998)
21. Wong, M.L., Mun, T.: Evolving recursive programs by using adaptive grammar based genetic programming. Genet. Program. Evolvable Mach. 6, 421–455 (2005)
22. Yu, T., Clark, C.: Recursion, lambda-abstractions and genetic programming. Cognitive Science Research Papers-University Of Birmingham CSRP, pp. 26–30 (1998)

Proposal and Preliminary Investigation of a Fitness Function for Partial Differential Models

Igor S. Peretta[1,2,3]($\boxtimes$), Keiji Yamanaka[1], Paul Bourgine[2,4], and Pierre Collet[2,3]

[1] Faculty of Electrical Engineering,
Federal University of Uberlandia, Uberlandia, Brazil
iperetta@ieee.org
[2] ECCE e-laboratory, CS-DC UNESCO UniTwin, Strasbourg, France
http://cs-dc.org
[3] ICUBE Laboratory, Strasbourg University, Strasbourg, France
[4] Faculty of Mathematics, Computing and Technology,
The Open University, Milton Keynes, UK

Abstract. This work proposes and presents a preliminary investigation of a fitness evaluation scheme supported by a proper genotype representation intended to guide an under development expansion to EASEA/EASEA-CLOUD platforms to evolve partial differential equations as models for a specific system of interest, starting with measures from that system. A simple proof of concept using a dynamic bidirectional surface wave is presented, showing that the proposed fitness evaluation scheme is very promising to enable automate system modelling, even when dealing with up to $\pm 10\%$ noise-added data.

Keywords: System modelling · Partial differential equations · Fitness function · Galerkin's method · Jacobi-Legendre polynomials · Tree-based Genetic Programming

1 Introduction

Systems modeling has important implications, from Physics and Chemistry to Biology and Social sciences [2]. Because many natural phenomena can be modelled in terms of differential equations, Genetic Programming could be used to perform symbolic regression in order to find the differential equations behind a data set obtained through observation. This is not a recent idea [6] but although symbolic regression is typically used to find explicit and differential equations, this research intends, as stated by [10], to detect any underlying physical law that the system of interest obeys to, rather than trying to model a specific signal. This is also explored by the work of [5] which presents a GP-based methodology to learn ordinary differential equations starting from experimental data.

In other words, this work is aimed to retrieve – with the help of an under development GP – underlying physical laws that could be described by a partial

© Springer International Publishing Switzerland 2015
P. Machado et al. (Eds.): EuroGP 2015, LNCS 9025, pp. 179–191, 2015.
DOI: 10.1007/978-3-319-16501-1_15

differential equation (PDE) from the data measured from the system of interest. Note that this is not symbolic regression: starting from measurements, the objective is to retrieve the PDE whose unknown function solution explains the data. Moreover, this work is part of a research to enable EASEA and EASEA-CLOUD platforms [1] to model complex systems using GP principles shown by [9].

Section 2 briefly describe some theoretical concepts related to approximating PDE solving. Section 3 presents developments and the proposed approach to evaluate fitness. Section 4 shows a practical example: using a multivariate model – a dynamic bidirectional surface wave with known solution – to simulate the measured data, along with fitness evaluation for seven arbitrary individuals. Section 5 draws conclusions and discusses ongoing work.

2   Theory

This proposal for fitness evaluation is based on the idea, among others, that it is possible to approximate PDEs solving. Some criteria to build those PDEs are presented in Sect. 3.5. A brief introduction is presented in this Section on weighted residual and Galerkin's method, as well as on Jacobi polynomials.

2.1   Weighted Residual and Galerkin's Method

The method presented by Galerkin [4] is widely classified into the class of spectral methods from the family of weighted residual methods. It could be defined as a numerical scheme to approximate the solving of differential equations represented by $\mathcal{D}\left[u(\boldsymbol{x})\right] = s(\boldsymbol{x})$.

Mostly, weighted residual methods are approximation techniques in which a residual $R\left[u(\boldsymbol{x})\right] = \mathcal{D}\left[u(\boldsymbol{x})\right] - s(\boldsymbol{x})$ (that represents the approximation error) is a quantity to be minimized ($R\left[u(\boldsymbol{x})\right] = 0$) [11], where $\boldsymbol{x}$ defines the domain of the problem; $\mathcal{D}$ is the functional known as the differential operator; $u(\boldsymbol{x})$ is the unknown solution; and $s(\boldsymbol{x})$, known as the source function, is independent of u.

Therefore, the differential equation is known to be presented in its residual form if it could be described as $\mathcal{D}\left[u(\boldsymbol{x})\right] - s(\boldsymbol{x})$ which is equal to zero.

An approximation $\hat{u}(\boldsymbol{x})$ to the unknown solution $u(\boldsymbol{x})$, also known as the trial function, is initially built as a projection on a function space characterized with a finite set of $N + 1$ basis functions $\mathcal{B} = \{\phi_i(\boldsymbol{x})\}|_{i=0}^{N}$, as shown in (1):

$$\hat{u}(\boldsymbol{x}) = \sum_{i=0}^{N} \tilde{u}_i \, \phi_i(\boldsymbol{x}) \tag{1}$$

where $\tilde{u}_i$'s are unknown coefficients of the trial function.

Weighted residual methods state that the minimized residual must be orthogonal to a set of arbitrary test functions. Galerkin's method presents the idea of using basis functions as test functions. It requires the PDE in its residual form (the residual) to be orthogonal to each of the early chosen basis functions in $\mathcal{B}$. Also, all those basis functions $\phi_i(\boldsymbol{x})$ must satisfy some previously known conditions (usually linear homogeneous boundary conditions) [7]. The approximate solution (trial function) is the truncated Galerkin expansion. To achieve this solution, a system of $N + 1$ equations is built using the orthogonality requirement, as shown in (2):

$$\langle \phi_n(\boldsymbol{x}), R\,[\hat{u}(\boldsymbol{x})]\,\rangle|_{n=0}^{N-1} = \langle \phi_n(\boldsymbol{x}), \mathcal{D}\,[\hat{u}(\boldsymbol{x})] - s(\boldsymbol{x})\rangle|_{n=0}^{N} = 0 \qquad (2)$$

where $\langle f(x), g(x) \rangle = \int_a^b f(x)g(x)w(x)\,dx$ is the inner product between $f(x)$ and $g(x)$ on the interval $[a, b]$ and $w(x)$ is a weight function which aids the definition of a Hilbert inner product space.

After solving (2) for $N+1$ unknown coefficients $\tilde{u}_i$ and plugging them into (1), the truncated Galerkin expansion for the differential equation solution is finally achieved. This approximate solution is the projection of the PDE solution on the Hilbert inner product space, *i.e.* a weighted sum of orthogonal functions.

2.2   Jacobi Polynomials

Using orthogonal polynomials with Galerkin's method ensures an orthogonal Hilbert space where any desirable smooth function could be projected, *i.e.* a powerful approximation could be built using truncated Galerkin expansions. Jacobi (orthogonal) polynomials are an interesting choice for basis functions due to some of their properties [7].

Jacobi polynomials have the univariate hypergeometric definition present in (3), as shown by [12] and [7].

$$P_n^{(\alpha,\beta)}(x) = \frac{\Gamma(n + \alpha + 1)}{\Gamma(n + 1)\gamma(\alpha + 1)}\,{}_2F_1\left(-n, n + \alpha + \beta + 1; \alpha + 1; \frac{1}{2}(1 - x)\right) \quad (3)$$

where $\Gamma(\cdot)$ is the gamma function; ${}_2F_1(p, q; r; z)$ is the Gauss's hypergeometric function with respect to z; $\alpha \geq -1$; $\beta \geq -1$; and $n \geq 0$ is the polynomial degree.

Using $\alpha \geq -1$ and $\beta \geq -1$ ensures the integrability of $w(x)$ [12]. Yet, when given the appropriate choice $\alpha = \beta$ the associated error is asymptotically minimized in an $L^{p(\alpha)}$-norm, as stated by [7]. Regarding Hilbert inner product space, Jacobi polynomials are orthogonal with respect to the interval $[-1, +1]$ and the weight function $w(x) = (1 - x)^\alpha (1 + x)^\beta$.

Special cases of Jacobi polynomials are achieved by choosing appropriate α and β. Basis functions could be generated to be asymptotically similar to Legendre polynomials, as adopted for this work, by choosing $\alpha = \beta = 0$.

When in need of arbitrary upper and lower limits for Jacobi polynomials, a linear mapping $\mathcal{M} : \{x \in \mathbb{R} \,|\, a \leq x \leq b\} \mapsto \{\xi \in \mathbb{R} \,|\, -1 \leq \xi \leq +1\}$ must be defined. Szëgo [12, pp. 58] proposed orthogonal polynomials with respect to the interval $[a, b]$ as defined in (4):

$$P_n^{(\alpha,\beta)}\left(2\frac{x - a}{b - a} - 1\right) = \frac{\Gamma(n + \alpha + 1)}{\Gamma(n + 1)\Gamma(\alpha + 1)}\,{}_2F_1\left(-n, n + \alpha + \beta + 1; \alpha + 1; \frac{b - x}{b - a}\right)$$

$$(4)$$

Differential operators deal with derivatives. An important identity for derivatives of Jacobi polynomials [7,8] is presented in (5):

$$\frac{d^k}{dx^k} P_n^{(\alpha,\beta)}(x) = \frac{\Gamma(n+\alpha+\beta+k+1)}{2^k \Gamma(n+\alpha+\beta+1)} P_{n-k}^{(\alpha+k,\beta+k)}(x) \tag{5}$$

Note that an arbitrary order k derivative of a Jacobi polynomial can be exchanged for another Jacobi polynomial, without any precision loss.

3 Proposed Method

3.1 Modifications to the Classical Galerkin Method

Starting from Szëgo's mapped Jacobi polynomials on the finite interval $[a, b]$ [12] and the derivation of identity (5) [8], this work was successful in achieving the derivation for the identity in (6):

$$\frac{d^k}{dx^k} P_n^{(\alpha,\beta)}\left(2\frac{x-a}{b-a}-1\right) = \frac{\Gamma(n+\alpha+\beta+k+1)}{(b-a)^k \Gamma(n+\alpha+\beta+1)} P_{n-k}^{(\alpha+k,\beta+k)}\left(2\frac{x-a}{b-a}-1\right) \tag{6}$$

This result, together with mapped polynomials in (4), enables to analytically operate differentials on Galerkin expansions which precedes definite integrations (inner products) from Galerkin's method. Note that both differential operator identity and definite integrations to be carried could be taken over arbitrary intervals. Those properties make mapped Jacobi polynomials an interesting option to work as basis functions when taking precautions on boundary conditions (e.g. [7]). Following this idea to build the basis set, the approximate solution has the form present in (7):

$$\hat{u}(x) = \sum_{i=0}^{N} \left[\tilde{u}_i \cdot P_i^{(\alpha,\beta)}\left(2\frac{x-a}{b-a}-1\right) \right] \tag{7}$$

where N is the polynomial degree of the truncated Galerkin expansion.

3.2 Multivariate Problems

When addressing to problems on multivariate domains, some adjustments must be done. One of them is the definition for the inner product which must be extended to support multiple integrals. Note that the use of Legendre polynomials ($\alpha = \beta = 0$) simplifies this effort, once $w(x) = 1$ whatever the adopted variable. The Galerkin system of equations, solved for the unknown coefficients of the expansion, is then built as shown in (8):

$$\langle \phi_n(\boldsymbol{x}), R[\hat{u}(\boldsymbol{x})] \rangle |_{n=0}^{(N+1)^d - 1} =$$

$$\int_{a_0}^{b_0} \int_{a_1}^{b_1} \cdots \int_{a_{d-1}}^{b_{d-1}} \phi_n(\boldsymbol{x}) \cdot \{ \mathcal{D}[\hat{u}(\boldsymbol{x})] - s(\boldsymbol{x}) \} \, dx_0 \, dx_1 \ldots dx_{d-1} \Bigg|_{n=0}^{(N+1)^d - 1} = 0 \tag{8}$$

where d is the number of dimensions (variables) of the domain of the problem; N is the predetermined polynomial degree with respect to all variables in the expansion; x is the vector of d variables $(x_0, x_1, \ldots x_{d-1})^T$; and a_i and b_i are the i-th lower and upper limits of integration, respectively, for $i = 0, \ldots d - 1$.

The truncated Galerkin expansion must be adjusted as in (9):

$$\hat{u}(x) = \sum_{i=0}^{(N+1)^d-1} \tilde{u}_i \, \phi_i(x) \tag{9}$$

where $\phi_i(x)$ is a multivariate basis function from a finite basis set with a span of $(N + 1)^d$ functions.

Finally, a proper basis set is built based on linear mappings $x_i \mapsto \xi_i$, i.e. mapping functions $\xi_i(x_i) = 2\frac{x_i - a_i}{b_i - a_i} - 1$ for $i = 0, \ldots d - 1$, and combinatorics of different Legendre polynomials, both with respect to each variable that constitutes the domain. This multivariate basis set is shown in (10):

$$\mathcal{B} = \{\phi_n(x)\}|_{n=0}^{(N+1)^d-1} = \left\{ \prod_{i=0}^{d-1} P_{n^\ddagger}^{(0,0)} (\xi_i(x_i)) \right\} \Bigg|_{n=0}^{(N+1)^d-1} \tag{10}$$

where $n^\ddagger = \left\{ \left\lfloor \frac{n}{(N+1)^i} \right\rfloor \mod (N+1) \right\}$ is the formula to keep track of the subindex related to the respective Legendre polynomial degree with respect to the i-th variable inside the product.

3.3 Building a Custom Galerkin System

This work successfully adopts a modification to the classic procedure for obtaining the Galerkin system: instead of getting all $(N+1)^d$ equations derived from the inner product statement, some equations could be coined from known information related to the problem (e.g. boundary conditions) applied to the truncated Galerkin expansion.

Once the value of the solution is known at a given point on the domain, i.e. both the measurement V and its respective domain coordinates $(x_0, x_1 \ldots x_{d-1})^T$ are available from the dataset, those numbers can be plugged to the Galerkin expansion in order to achieve a complementary equation to the system in the form of (9) and (10), i.e. $\hat{u}((x_0, x_1 \ldots x_{d-1})^T) = V$. A minimum of 2^d points is required to this approach, which means that 2^d complementary equations could be written. Note that, in this way, basis functions themselves do not need to satisfy boundary conditions which are already part of the system of equations.

Together with the first $(N + 1)^d - 2^d$ equations produced by inner products, a "custom" Galerkin system is then built, as seen in (11):

$$\begin{cases} \int_{a_0}^{b_0} \int_{a_1}^{b_1} \cdots \int_{a_{d-1}}^{b_{d-1}} \phi_n(x) \cdot \{\mathcal{D}\left[\hat{u}(x)\right] - s(x)\} \, dx_0 \, dx_1 \ldots dx_{d-1} \Bigg|_{n=0}^{(N+1)^d-2^d-1} = 0 \\[2em] \sum_{m=0}^{(N+1)^d-1} \tilde{u}_i \cdot \left[\prod_{i=0}^{d-1} P_{\left\{ \left\lfloor \frac{m}{(N+1)^i} \right\rfloor \mod (N+1)\right\}}^{(0,0)} (\xi_i(X_{ni})) \right] - V_n \Bigg|_{n=0}^{2^d-1} = 0 \end{cases} \tag{11}$$

where $\boldsymbol{X}_n = (X_{n0}, X_{n1}, \ldots X_{nd})^T$ is the point on the domain at which the solution is known to assume value V_n.

Note that this custom system could be a linear or a non-linear one, depending only on the differential operator which is part of the partial differential equation in its residual form. Also, regarding this work, it is important to notice that Galerkin method approximation quality depends on the chosen polynomial degree N which needs to be chosen beforehand.

3.4 Data Preparation

After collecting experimental data from the system of interest, the preparation step takes place before the GP run and consists in dividing the domain into overlapping sub-domains[1] defined by $2^d + 1$ points which are the closest possible to each other and have non-zero depth in all related dimensions. Each of those sub-domains has 2^d points necessary to coin complementary equations to the custom Galerkin system in (11) and an extra point to be used as a reference to

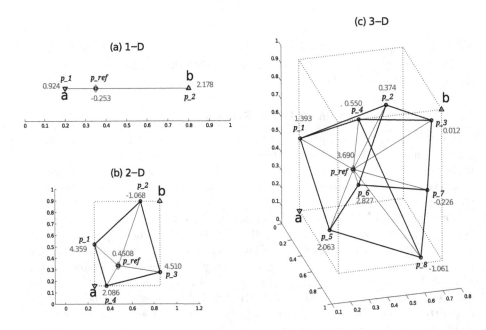

Fig. 1. Examples of sub-domain $\mathcal{S}$ with a total of $2^d + 1$ points in one, two or three dimensions. Legend: $\circ$ represents a point from the database with known coordinates and respective value for the quantity of interest; $\Diamond$ is the reference point; ∇ and $\triangle$ are calculated *a posteriori* as the minimum coordinates $\boldsymbol{a}$ and maximum coordinates $\boldsymbol{b}$, respectively. Different from other points, $\boldsymbol{a}$ and $\boldsymbol{b}$ do not have a known value for the quantity of interest as they only define limits of integration.

---

[1] The number of sub-domains has as an upper bound the number of available points in the dataset.

evaluate fitness. This reference point must be the closest one to the sub-domain "center of mass", in order to raise the quality of evaluations. The idea is to verify if the approximate solution of the differential could explain measurements from the dataset. Figure 1 shows examples of 1, 2 or 3-dimensional sub-domains.

Points a and b are calculated to define limits of integration for inner products and to verify if their respective sub-domain has non-zero depths in all dimensions. The set of sub-domains is identified by $\mathcal{S} = \{\mathcal{S}_i\}$. Each i-th sub-domain, therefore named as $\mathcal{S}_i$, is defined in this work by having four members: P_{2^d} which is a list of 2^d points from the domain; V_{2^d}, their respective measured values for the quantity of interest; P_{ref}, the point of reference; and V_{ref}, the value for the quantity of interest at the reference.

This preparation step is essential to the proposed fitness function, but it needs to be performed just once to be used as input data to the GP run.

3.5   Representing PDEs Within Genetic Programming

To the system modelling GP aimed in this research, each individual is a PDE, *i.e.* a candidate model for the system of interest, in its residual form. The aim is to evolve those differentials until the emergence of a model whose solution could explain the measure data. An algebraic syntax tree representation was chosen as the genotype to enable heuristics to manipulate individuals and a simplified compiler evaluator to take place during execution time. Figure 2 shows an example that could be evolved in this scenario.

Those syntax trees could contain one to several of the following nodes:

- Constants: a predetermined finite set of arbitrary real numbers;
- Variables: user can decide the number of dimensions for the domain, they will be identified by sub-index for the x symbol (e.g. x_0, x_1);
- Operator: the four arithmetic operators – plus, minus, division, and product;
- Function: trigonometric, mathematical or user defined ones;

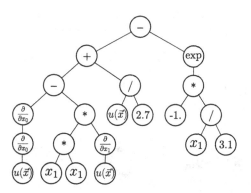

Fig. 2. Algebraic syntax tree representing the following partial differential equation in its residual form: $\frac{\partial^2}{\partial x^2}u(x,y) - y^2\frac{\partial}{\partial y}u(x,y) + \frac{u(x,y)}{2.7} - \exp\left(-\frac{y}{3.1}\right) = 0$.

– Unknown function: this special node represents the multivariate function (with respect to all related variables) that is supposed to explain the measure data, i.e. this is the solution of the desired partial differential model;
– Derivative of first order: this special node represents the derivative which can operate the unknown function or another derivative (to build up higher order derivatives) – the limitation of not operating any other node could be supported by the product rule.

3.6   Fitness Evaluation

The present proposal for fitness evaluation of partial differential individuals could be resumed by the following procedure:

1. Take a given individual (Sect. 3.5) and apply Galerkin method (Sect. 3.3) using boundary conditions given by known points for each predefined sub-domain (Sect. 3.4) to achieve piecewise approximate solutions;
2. For each sub-domain, plug the related reference point (the "plus one") to the respective approximate solution and an estimated value for the quantity of interest at that point is evaluated;
3. Use a metric to keep track of errors between estimated values from approximate solutions and registered values from measure data – in this work, the squared error was used to get the *mean squared error* (MSE);
4. After evaluating the error on every valid sub-domain, the expected value of those errors (MSE) is taken as the fitness for the evaluated individual, which turn this into a minimization problem.

4   Results/Discussion

The present case study assumes as the physical system of interest a *dynamic bidirectional surface wave*. This problem is defined in (12):

$$\begin{cases} \text{PDE}: k_0 \frac{\partial}{\partial x} u(x,y,t) + k_1 \frac{\partial}{\partial y} u(x,y,t) + k_2 \frac{\partial}{\partial t} u(x,y,t) + k_3\, u(x,y,t) = 0 \\ \text{IC}: \qquad\qquad\qquad u(x,y,0) = e^{-\left(x^2+y^2\right)} \end{cases}$$

$$(12)$$

where k_0, k_1, k_2, and k_3 are known constants.

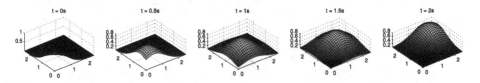

Fig. 3. Example for a dynamic bidirectional surface wave simulation over the direction domain $0 \leq x, y \leq 2.5\,\text{m}$ and time domain $0 \leq t \leq 2\,\text{s}$.

The known solution for (12) is $u(x,y,t) = e^{-\left[\left(x-\frac{k_0 t}{k_2}\right)^2 + \left(y-\frac{k_1 t}{k_2}\right)^2\right] - \frac{k_3 t}{k_2}}$ (adapted from [3, pp. 211]). In Fig. 3, a simulation with $k_0 = k_1 = k_2 = 1$ and $k_3 = 0.1$.

4.1 "Sampling" the Simulated Data

In order to have available measure data to perform tests, a surface wave was simulated with arbitrary constants (Fig. 3) based on the known solution of (12). For each time $t \in \{0., 0.25, 0.5, 0.75, 1., 1.25, 1.5, 1.75, 2.\}$ there were 30 two-dimensional points generated at random (for spatial dimensions x and y). Those 270 three-dimensional points had their "measures" defined by the known solution equation. After, the preparation step (Sect. 3.4) achieved a set of 157 sub-domains with 9 points each $(2^d + 1,$ with $d = 3)$.

The first experimental run is referred in this work as the *noise-free experiment*. Table 1 shows some statistics of the data. This work also uses other "measure" datasets, both developed by applying noise to "measures" as in (13).

$$V_{\text{noise}} = V_{\text{original}} \cdot \left[1 + \mathcal{U}(-l, +l)\right], \tag{13}$$

where V_{original} are the original values; $\mathcal{U}(-l, +l)$ is an uniform random variable ranging from $-l$ to $+l$; and V_{noise} are the noise-added values. Note that coordinates from the dataset are kept the same. The noise must be added before the preparation step due to the overlapping of sub-domains. After the preparation step, those points have formed the exact same set of 157 sub-domains, but with updated noisy values. Second and third experimental runs are referred as *noise-add experiments* and are related to their noise level l.

Table 1. Simulated "measure" data statistics, noise-free.

Minimum	Q1	Median	Q3	Maximum
$2.066 \ 10^{-3}$	$105.475 \ 10^{-3}$	$295.814 \ 10^{-3}$	$542.981 \ 10^{-3}$	$947.308 \ 10^{-3}$

4.2 Applying Fitness to Arbitrary Individuals

Here are some results of the proposed fitness evaluation scheme when applied to candidate models for the dynamic bidirectional surface wave. Table 2 shows seven arbitrary individuals coined to this example that could be evolved by GP.

Note that PDE-3 is the "optimal" which GP needs to reach. PDE-2 is practically the same, differing just by the last term coefficient. PDE-1 is slightly different due to a 2nd-order derivative and the last term coefficient. PDE-4 and PDE-5 are structurally different from PDE-3. Finally, PDE-6 and PDE-7 are listed cause their developments lead to ill-posed Galerkin system of equations.

Noise-free experiment: Table 3 shows resulting fitness evaluation for individuals from Table 2. Individual PDE-3 has the lowest fitness, being the best individual so far as expected, followed by PDE-2.

Table 2. Individuals and respective residual form PDEs. The symbol u denotes the function $u(x, y, t)$; u_x is the first order partial derivative of $u(x, y, t)$ with respect to x; u_{xx} is the respective second order with respect to x, and so on.

Individual	PDE, residual form
PDE-1	$u_{xx} - u_y + u_t - 0.3\,u$
PDE-2	$u_x + u_y + u_t + 0.2\,u$
PDE-3	$u_x + u_y + u_t + 0.1\,u$
PDE-4	$u_{xx} - 3\,u_{xy} + u_{yy} - 0.2\,u_t + 0.1\,u$
PDE-5	$-3\,u_{xy} + 2\,u_t - 5\,u$
PDE-6	$u_t + x\,y\,t$
PDE-7	$10\,u_t + x\,y$

Noise-add experiment, level 5 %: Table 4 shows fitness when original measure data is updated according (13) with noise level $l = 0.05$. Note that individual PDE-3 continues to have the lowest fitness, followed by PDE-2. Fitness of PDE-5 has changed dramatically within this scenario.

Noise-add experiment, level 10 %: Table 5 shows fitness when original measure data is updated according (13) with noise level $l = 0.1$. Individuals PDE-3 and PDE-2 continues to lead. PDE-5 fitness followed the former scenario.

This case study indicates that the proposed fitness evaluation scheme for partial differential individuals was successful to point out fit individuals to the problem, which means it could guide evolution of models within a GP run. Even when noisy data is available (up to $\pm 10\,\%$), the proposed scheme could indicate the right model (PDE-3) to the case study, a dynamic bidirectional surface wave.

Regarding other individuals, PDE-2 also has its fitness evaluated in agreement of what should be expected, as long as it is the closer to PDE-3 of all others. PDE-5 and PDE-6 are the worst fit of all, even that PDE-5 has improved its

Table 3. Comparative table for fitness. Each PDE has an approximate polynomial solution built with: 64 terms; with respect to the three variables of the problem; and of polynomial degree 3. Noise-free data. Valid sets refers to non-improper evaluations.

Individual	Fitness (MSE)	Valid sets
PDE-1	$35.542\,10^{-3}$	99.4 %
PDE-2	$2.759\,10^{-3}$	96.8 %
PDE-3	$1.402\,10^{-3}$	100.0 %
PDE-4	$11.889\,10^{-3}$	95.5 %
PDE-5	$971.027\,10^{-3}$	97.5 %
PDE-6	$834.214\,10^{-3}$	100.0 %
PDE-7	$91.605\,10^{-3}$	100.0 %

Table 4. Comparative table for fitness. Approximate solutions are three-dimensional polynomials with 64 terms and degree 3. Performed on $\pm 5\%$ noise-add data.

Individual	Fitness (MSE)	Valid sets
PDE-1	$55.787\,10^{-3}$	99.4%
PDE-2	$6.378\,10^{-3}$	96.2%
PDE-3	$4.977\,10^{-3}$	100.0%
PDE-4	$15.951\,10^{-3}$	94.9%
PDE-5	$116.398\,10^{-3}$	96.8%
PDE-6	$818.794\,10^{-3}$	100.0%
PDE-7	$86.113\,10^{-3}$	100.0%

Table 5. Comparative table for fitness. Approximate solutions are three-dimensional polynomials with 64 terms and degree 3. Performed on $\pm 10\%$ noise-add data.

Individual	Fitness (MSE)	Valid sets
PDE-1	$94.112\,10^{-3}$	99.4%
PDE-2	$18.105\,10^{-3}$	96.2%
PDE-3	$16.817\,10^{-3}$	100.0%
PDE-4	$22.059\,10^{-3}$	93.6%
PDE-5	$122.419\,10^{-3}$	96.8%
PDE-6	$804.932\,10^{-3}$	100.0%
PDE-7	$82.178\,10^{-3}$	100.0%

fitness when dealing with noisy data. All of those could benefit from the used linear solver, as far as all PDEs achieved more than 93% valid solutions for the 157 possible sub-domains, even with noisy data. Note that the scheme is able to evaluate PDE-6 and PDE-7 with their ill-posed Galerkin system of equations. The order of magnitude for listed fitnesses must be analysed in the light of Table 1 which shows the magnitude of measure data.

5   Conclusion and Developments

In this work, a proposal for fitness evaluation of partial differential individuals supported by a proper genotype representation is presented. This scheme to evaluate fitness was able to deal with the case study, a dynamic bidirectional surface wave, achieving promising results that could fulfil expectations for a robust and versatile automated system modelling application, even when dealing with up to $\pm 10\%$ noise-added data. Some advantages must be pointed out:

– This approach is relatively robust due to local evaluations, *i.e.* possible local approximation errors could be minimized when consolidating the whole;

- Possible improper calculations do not invalidate fitness evaluation, *i.e.* if some sub-domain contains improper data that could turn into a numerical calculation exception, the scheme just dismisses it without greater losses;
- Due to arithmetic and algebraic manipulation, as with identities in (4) and (5), procedures for building equations are not subject to precision losses or other approximation issues – the scheme itself is able to be solved analytically – only that solutions to those equations are approximations;
- In theory, it can work with several dimensions (scalable property), limited only by the required computational power;
- In this scheme, "time" is considered to be just another dimension.

Also, some issues that are subject of further investigation are the following:

- High computational execution time (few hours per individual) – authors are currently adapting algorithms to CUDA C/C++ to integrate to EASEA and EASEA-CLOUD platforms which can provide easy massive parallelization of evolutionary algorithms on GPGPUs for improved performance [1].
- The choice for polynomial degree of the truncate Galerkin expansion must rely on data sufficiency assumption, which is subjective, especially in higher dimensions – one workaround is to ensure sufficient measurements from the system of interest, enough to keep polynomial degree between three or four.
- As the idea is that GP must evolve a model that reflects the underlying law that the system of interest obeys and the Galerkin method assumes all expansion coefficients equal to zero if the residual does not contain terms with derivatives, GP must ensure individuals being differential equations.

Developments: Currently, the scheme is implemented and dealing only with linear differential individuals due to the adopted solver for the system of equations. Theoretically, this very scheme could also work with non-linear individuals. Authors are researching possibilities for non-linear solvers that must deal with large systems of equations that can deal with eventual ill-posed systems.

As the main objective is to perform system modelling starting from measure data, the GP application needs to work with the three classical subsets (training, validation and test) to avoid over-fitting. The initial idea is to get the set of sub-domains to be randomly divided into three subsets. For example, the 157 from the case study would turn into three subsets with ≈ 52 sub-domains each.

Another step into the research is to enable this complex system modelling GP to handle not only one, but a set of PDEs. When dealing with scalar fields, it is usual to deal with a model which is a single PDE. When starting to deal with vector fields, though, it becomes necessary to deal with sets of PDEs.

Acknowledgements. I. S. Peretta would like to thank the non-simultaneous support received from CAPES (PDSE scholarship #18386-12-1) and CNPq (Full PhD scholarship - GD), both Brazilian funding agencies.

References

1. Collet, P., Krüger, F., Maitre, O.: Automatic parallelization of EC on GPGPUs and clusters of GPGPU machines with EASEA and EASEA-CLOUD. In: Tsutsui, S., Collet, P. (eds.) Massively Parallel Evolutionary Computation on GPGPUs. Natural Computing Series, pp. 35–61. Springer, Heidelberg (2013)
2. Drazin, P.G.: Nonlinear Systems. Cambridge Texts in Applied Mathematics, Cambridge (1997)
3. Farlow, S.J.: Partial Differential Equations for Scientists and Engineers. Dover Publications, New York (1993)
4. Galerkin, B.G.: Series occuring in various questions concerning the elastic equilibrium of rods and plates. Eng. Bull. (Vestnik Inzhenerov) **19**, 897–908 (1915). (in Russian)
5. Gaucel, S., Keijzer, M., Lutton, E., Tonda, A.: Learning dynamical systems using standard symbolic regression. In: Nicolau, M., Krawiec, K., Heywood, M.I., Castelli, M., García-Sánchez, P., Merelo, J.J., Rivas Santos, V.M., Sim, K. (eds.) EuroGP 2014. LNCS, vol. 8599, pp. 25–36. Springer, Heidelberg (2014)
6. Koza, J.R.: Genetic Programming: On the Programming of Computers by Means of Natural Selection. MIT Press, Cambridge (1992)
7. Livermore, P.W., Ierley, G.R.: Quasi-Lp norm orthogonal Galerkin expansions in sums of Jacobi polynomials. Numerical Algorithms **54**(4), 533–569 (2010)
8. Luke, Y.L.: The Special Functions and their Approximations. Academic Press, New York (1969)
9. Maitre, O.: Genetic programming on GPGPU cards using EASEA. Massively Parallel Evolutionary Computation on GPGPUs. Natural Computing Series, pp. 227–248. Springer, Heidelberg (2013)
10. Schmidt, M., Lipson, H.: Distilling free-form natural laws from experimental data. Science **324**, 81–85 (2009)
11. Shen, J., Tang, T., Wang, L.L.: Spectral Methods: Algorithms, Analysis and Applications. Springer, Heidelberg (2011)
12. Szegö, G.: Orthogonal Polynomials, American Mathematical Society Colloquium Publications, American Mathematical Society, vol. 23, revised edition (1959)

Evolutionary Methods for the Construction of Cryptographic Boolean Functions

Stjepan Picek[1,3], Domagoj Jakobovic[1(✉)], Julian F. Miller[2], Elena Marchiori[3], and Lejla Batina[3]

[1] Faculty of Electrical Engineering and Computing,
University of Zagreb, Zagreb, Croatia
stjepan@computer.org, domagoj.jakobovic@fer.hr
[2] Department of Electronics, University of York, York, UK
[3] Radboud University Nijmegen, Nijmegen, The Netherlands

Abstract. Boolean functions represent an important primitive when constructing many stream ciphers. Since they are often the only non-linear element of such ciphers, without them the algorithm would be trivial to break. Therefore, it is not surprising there exist a substantial body of work on the methods of constructing Boolean functions. Among those methods, evolutionary computation (EC) techniques play a significant role. Previous works show it is possible to use EC methods to generate high-quality Boolean functions that even surpass those built by algebraic constructions. However, up to now, there was no work investigating the use of Cartesian Genetic Programming (CGP) for producing Boolean functions suitable for cryptography. In this paper we compare Genetic Programming (GP) and CGP algorithms in order to reach the conclusion which algorithm is better suited to evolve Boolean functions suitable for cryptographic usage. Our experiments show that CGP performs much better than the GP when the goal is obtaining as high as possible nonlinearity. Our results indicate that CGP should be further tested with different fitness objectives in order to check the boundaries of its performance.

Keywords: Boolean functions · Genetic programming · Cartesian Genetic Programming · Cryptographic properties · Comparison

1 Introduction

Most cryptographic systems in use today are built as hybrid cryptosystems. In these systems asymmetric-key cryptography is used to exchange the keys and symmetric-key cryptography is used to encrypt and decrypt data. This separation is due to the fact that symmetric-key cryptography is much faster than the asymmetric-key [8]. The name symmetric-key denotes the fact that the same key is used for both data encryption and decryption.

One usual division of symmetric-key cryptography is block and stream ciphers [8]. In block ciphers algorithms encrypt and decrypt data in blocks of

© Springer International Publishing Switzerland 2015
P. Machado et al. (Eds.): EuroGP 2015, LNCS 9025, pp. 192–204, 2015.
DOI: 10.1007/978-3-319-16501-1_16

certain size and in stream ciphers this is done bitwise. In both of these types of cipher often the only nonlinear elements are Boolean functions or vectorial Boolean functions (vectorial Boolean functions are better known as Substitution boxes or S-boxes). Boolean functions are in general used in stream ciphers whereas S-boxes are used in block ciphers. In the rest of this paper we concentrate only on Boolean functions suitable for cryptographic usage in stream ciphers.

There exist three main approaches to generate Boolean functions for cryptographic usage: algebraic construction, random generation and heuristic construction [6]. In algebraic construction one usually uses some mathematical procedure that gives very good results such as the cipher RAKAPOSHI [3]. One of the most famous constructions is a finite field inversion [17]. However, although finite field inversion can be used to generate S-boxes with the highest possible nonlinearity levels, this is not so for Boolean functions. Furthermore, such constructions cannot give optimal results when considering side-channel attack resistance [21].

Random generation of Boolean functions has its strong points, the most prominent being that it is easy and fast, but the resulting Boolean functions usually have suboptimal properties for cryptography [11].

Heuristic methods offer easy and efficient way of producing large number of Boolean functions with very good cryptographic properties [2]. Among other heuristic methods, evolutionary computation (EC) offers highly competitive results when generating Boolean functions [19]. More details about different methods for evolving Boolean functions are given in Sect. 1.1. However, as far as the authors know, Cartesian Genetic Programming (CGP) has never been used for constructing Boolean functions suitable for cryptography. Since CGP is recognized as a suitable option when generating Boolean functions [13,14], its absence in the evolution of cryptography-suitable Boolean function creation is somewhat surprising.

In this paper we concentrate only on Boolean functions with 8 inputs since that represents the size most used in practical scenarios (e.g. cipher RAKAPOSHI [3]). Evolving Boolean functions with 8 inputs is a challenging task since there exist 2^{2^n} possible functions of n inputs (i.e. for 8 inputs this gives 2^{256} candidate solutions). To serve as a benchmark problem when comparing the algorithms, we look for a balanced Boolean function with maximum nonlinearity. However, this problem should not be only be considered as a benchmark, but rather as a difficult problem that has practical implications. It is a well known fact among the cryptography community that the upper bound for the nonlinearity property in the case of an 8-bit balanced Boolean function is 118 [22]. However, no one has been able to find such a function. Indeed, finding it would represent not only a significant result from the cryptographic perspective but also from the EC perspective since it would help profile EC methods as the truly viable option for cryptographic usages.

1.1 Related Work

As previously stated, there have been several applications of heuristic methods to the generation of Boolean functions for cryptography. Here we give only a few representative examples of work related to our research.

Millan et al. use Genetic Algorithms to evolve Boolean functions that have high nonlinearity [10]. Clark et al. experiment with Simulated Annealing when evolving Boolean functions with cryptography-relevant properties [4]. Burnett in her thesis use Genetic Algorithms in a combination with hill climbing to evolve Boolean functions and S-boxes [2]. McLaughlin and Clark on the other hand use Simulate Annealing to evolve Boolean functions that have several cryptographic properties with optimal values [9]. Picek et al. experiment with Genetic Programming and Genetic Algorithms to find Boolean functions that possess several optimal properties [19]. Several evolutionary algorithm methods are used by Picek et al. to evolve Boolean functions that have better DPA-related properties [18]. With the goal of finding maximal nonlinearity values of Boolean functions Picek et al. experiment with a handful of evolutionary algorithms and approaches [20]. Hrbacek and Dvorak use CGP to evolve bent Boolean functions of size up to 16 inputs. However, since bent Boolean functions should not be used in cryptography [5] this work has a limited applicability from the cryptographic perspective.

1.2 Our Contributions

To our best knowledge we are the first to consider CGP when evolving Boolean functions suitable for cryptographic usage. Furthermore, we experiment with different genotype sizes and mutation rates to investigate their influence on the ability of CGP to find good solutions. Since there is no prior experimental work, this should also be regarded as a guideline for future research. When experimenting with GP, we also investigate the influence of tree depth on the quality of the obtained solutions. We compare GP and CGP algorithms on a real-world difficult cryptographic problem to investigate their suitability.

The remainder of this paper is organized as follows: in Sect. 2 we describe relevant cryptographic properties and representations of Boolean functions. In Sect. 3 experimental setup and algorithms are presented, while results and short discussion are given in Sect. 4. Finally, Sect. 5 concludes with some suggestions for future work.

2 Boolean Functions and Their Properties

In this section we give a short overview of relevant cryptographic properties of Boolean functions. For further details we refer interested readers to [1,5].

The inner product of vectors a and b is denoted as $a \cdot b$. It is defined as $\oplus_{i=1}^{n} a_i b_i$, where "$\oplus$" represents addition modulo 2 (bitwise XOR).

A Boolean function f on $\mathbb{F}_2^n$ can be uniquely represented by a truth table (TT), which is a vector $(f(\mathbf{0}), ..., f(\mathbf{1}))$ that contains the function values of f, ordered lexicographically [1].

The Hamming weight $HW(f)$ of a Boolean function f is the number of ones in its binary truth table representation [1].

The second unique representation of Boolean function is the Walsh transform. It measures the similarity between $f(\mathbf{x})$ and the linear function $\mathbf{a} \cdot \mathbf{x}$ [1]. The Walsh transform of a Boolean functions f equals:

$$W_F(\mathbf{a}) = \sum_{x \in \mathbb{F}_2^n} (-1)^{f(\mathbf{x}) \oplus \mathbf{a} \cdot \mathbf{x}}. \tag{1}$$

A Boolean function is **balanced** (denoted "BAL" throughout the paper) if its Hamming weight is equal to 2^{n-1} [1].

The **nonlinearity** NL_f of a Boolean function f can be expressed in terms of the Walsh coefficients as [1]:

$$NL_f = 2^{n-1} - \frac{1}{2} max_{a \in \mathbb{F}_2^n} |W_f(\mathbf{a})|. \tag{2}$$

A Boolean function f is **correlation immune** of order t - $CI(t)$ if the output of the function is statistically independent of the combination of any t of its inputs [1]. For the Walsh spectrum it holds that

$$W_f(\mathbf{a}) = 0, \text{ for } 0 \leq HW(\mathbf{a}) \leq t. \tag{3}$$

A Boolean function f is **t-resilient** if it is balanced and with correlation immunity of degree t [1]. Due to the lack of space, we do not explain the roles of each property in the security application of Boolean function, but we rather refer readers to relevant literature.

2.1   Balanced Boolean Functions and Maximal Nonlinearity

Sarkar and Maitra showed that if a t-resilient Boolean function f has an even number of inputs n and $t + 1 \leq \frac{n}{2} - 1$ then its nonlinearity NL_f is bounded as follows [22]:

$$NL_f \leq 2^{n-1} - 2^{\frac{n}{2}-1} - 2^{t+1}. \tag{4}$$

Since we are looking for a Boolean function that has maximal nonlinearity, we can see that the resilience needs to be 0 which then simplifies the equation to the following one:

$$NL_f \leq 2^{n-1} - 2^{\frac{n}{2}-1} - 2. \tag{5}$$

From the formula it follows that the maximum nonlinearity for $n = 8$ and $t = 0$ equals 118.

3 Algorithms and Experimental Setup

We remind the reader that we focus on the evolution of Boolean functions that are balanced and with as high nonlinearity as possible. Naturally, the end goal is to find such a function that has nonlinearity 118, but even lower values can help us to reach the conclusion about the strength of a certain method. Moreover, such Boolean functions can have also practical applications in the design of stream ciphers. To conclude, the goals of our experiments can be stated through the following questions.

- Is CGP suitable for evolving Boolean functions when the focus is on the cryptographic usage?
- What is the influence of the genotype size on the quality of the solutions obtained?
- How does the performance of CGP compare with GP?
- What is the influence of tree depth in GP when evolving cryptographically suitable Boolean functions?

Additionally, we experiment with Genetic Algorithm (GA) which serves as a basic case scenario to determine a reference performance of the algorithm.

3.1 Genetic Algorithm

Our GA implementation uses the function truth table as chromosome representation, which is an array of bits of length 2^n, where n is the size of a Boolean function (therefore, in this research the chromosome length is 256 bits). For GA we use a steady state tournament selection with tournament size k equal to 3 and population size 100. In steady state tournament selection mechanism the worst of k randomly selected individuals is identified and replaced with a new individual. The new individual is constructed with the crossover of two random surviving parents from the tournament. After crossover, each new individual undergoes a mutation with a given probability.

We experimented with several genetic operators, but the best results were obtained with one-point crossover and simple mutation which inverts a randomly selected bit.

3.2 Tree-Based Genetic Programming

In Genetic Programming a function is represented as a tree of a certain depth. The inner nodes (function set) of a tree are Boolean primitives (such as AND, OR, NOT), while the leaves (terminals) may be a single input Boolean variables (v0..v7). We use the same function set, which is given below, for both GP and CGP. In GP experiments, the mutation probability is set to 0.3 per individual, and the population size is 500. Steady-state tournament selection with tournament size of 3 is used.

A small number of experiments were also conducted to select the appropriate operators, and based on that we used a simple tree crossover with 90 % bias for functional nodes and a subtree mutation.

The maximum tree depth is a parameter that is selected by the user and influences the available genotype size. When GP/CGP is used, one is effectively evolving a digital circuit and then examining its truth table to assess whether the function has the desired properties (e.g. balancedness or nonlinearity). However, with a GA approach one is directly evolving a truth table, so that the question of how it is implemented is not involved. Indeed the size of the truth table determines the size of the GA genotype (bitstring) whereas in the GP/CGP approaches, the size of the genotype is not directly related to the size of the desired truth table.

3.3   Cartesian Genetic Programming

In Cartesian Genetic Programming a program is represented as an indexed graph. The graph is encoded in the form of a linear string of integers [15]. Terminal set (inputs) and node outputs are numbered sequentially. Node functions are also numbered separately [15].

CGP has three parameters that are chosen by the user; number of rows n_r, number of columns n_c and levels-back l [14]. The number of rows and number of columns make the two-dimensional grid of computational nodes and their product gives the maximum number of computational nodes. The levels-back parameter controls the connectivity of the graph, i.e. it determines which columns a node can get its input from [14].

In CGP the genotype is a list of integers that represents the program primitives and how they are connected together [16]. The genotype is mapped to the directed graph that is executed as a program. Genotypes are of fixed length while phenotypes have variable length in accordance with the number of unexpressed genes.

The maximal length of the genotype is given by the following formula:

$$max\_length = n_r n_c (n_n + l) + n_o. \qquad (6)$$

In this application the number of node input connections n_n is 2 and the number of program output connections n_o is 1. The population size for CGP equals 5 in all our experiments. For CGP individual selection we use a $(1 + 4)$-ES evolution strategy in which offspring are favored over parents when they have a fitness less than or equal to the fitness of the parent. The mutation operator is one-point mutation where the mutation point is chosen with a fixed probability. The number of genes mutated is defined as fixed percentage of the total number of genes. Note, the single output gene is not mutated and is taken from the last node in the genotype. The genes chosen for mutation might be a node input connection or a function. For more details about CGP we refer readers to [13–16].

3.4   Fitness Functions

When searching for a balanced function with the best possible nonlinearity, we experimented with two fitness functions, both to be maximized. The first fitness function simply adds the balancedness penalty and nonlinearity values.

$$fitness = BAL + NL_f. \tag{7}$$

When a Boolean function is balanced we assign the BAL component a value of 0, and when it is unbalanced we assign it the negative difference up to the balancedness (i.e. the number of bits that need to be changed to reach balancedness) multiplied with a constant c. Based on the results from [19,20] we set that constant to 5 so that the unbalancedness penalty exceeds the values of nonlinearity.

For the second fitness function, we have used a two stage fitness in which a fitness bonus equal to the nonlinearity is awarded only to a genotype that is perfectly balanced (this occurs when $BAL = 0$); otherwise, the fitness is only the balancedness penalty. This is given in Eq. 8. The delta function $\delta_{BAL,0}$ takes the value one when $BAL = 0$ and is zero otherwise.

$$fitness = BAL + \delta_{BAL,0}NL_f. \tag{8}$$

Two stage fitness functions are commonly used in CGP when one is trying to optimize one quantity under a strict constraint; for instance, when trying to evolve a Boolean function that exactly matches a given truth table but which has the minimum number of gates [7]. Note that when Eq. 8 is used, one does not have to assign weights to the relative importance of different objectives. In Eq. 7 a nearly balanced Boolean function with high nonlinearity could receive the same fitness score as a fully balanced Boolean function with a lower nonlinearity. In the two stage fitness function described in Eq. 8 unbalanced Boolean functions are not assessed for nonlinearity at all. Note that we do not use a multiobjective approach, since the balancedness is a constraint rather than a separate objective.

An observant reader can easily notice that in Eq. 4 there is a resilience term which we know needs to be 0 so we disregard it and proceed to Eq. 5. The question is, should we disregard this property so readily? It is clear from those two formulas that the nonlinearity property changes in jumps of two and it always has an even value for Boolean functions with even number of inputs.

This means, if we reach the nonlinearity of 116, to move to the value of 118 actually a random search is performed - since there are no values between those two, the evolutionary algorithm has no means of differentiating different solutions with nonlinearity 116. To add this missing information, we may include the resilience property in the fitness function.

However, the problem is that we do not know what resilience values can lead to nonlinearity 118. It is plausible to consider it better to have the resilience as small as possible, since we know that for the best nonlinearity the resilience must be 0. However, it is possible that Boolean functions with resilience larger than 0 can lead the search towards new, unexplored areas which can eventually lead to nonlinearity 118. Since there is no research investigating those conditions at this moment, all that researchers can do is use their intuition to decide on the best approach. We take into account the first option where we add to the fitness function the constraint that the resilience must be 0 and carry out empirical experiments.

3.5   Experimental Setup

Since there are no previous results when using CGP to evolve Boolean functions with good cryptographic properties, first we need to consider how to set CGP parameters. Setting the number of rows to be 1 and levels-back to be equal to the number of columns is regarded as the best and most general choice [14]. This choice should be used when there is no specialist knowledge about the problem.

However, this still leaves open the question what should be the size of the number of columns parameter. Furthermore, CGP usually uses small population sizes and has no crossover operator [14]. Determining the best combination of maximum number of nodes (gates in this case) and mutation rate is an important step in hitting the parameter sweet spot for CGP. Indeed, it has been shown that generally very large genotypes and small mutation rates perform very well [12]. Thus some experiments were performed varying these two parameters.

Common Parameters. The following parameters of the experiments are common for all algorithms: the size of Boolean function is 8 (the size of the truth table is 256) and the number of independent runs for each experiment is 50. The function set n_f for both GP and CGP in all the experiments consists of binary Boolean primitives OR, XOR, AND, XNOR and AND with one input inverted. For the stopping condition we use the number of evaluations which we set to 500 000.

4   Results and Discussion

First we note that for the GA case, the best obtained result are balanced functions with nonlinearity value of 112 with the average of 111.8 over 50 runs. This is considerably worse than the best (and most average) solutions obtained with CGP, as shown below.

Furthermore, in all the experiments so far, we have been unable to obtain the nonlinearity of 118; only the value of 116 could be found for balanced functions. While not the maximum, this nonlinearity level is still very high for practical purposes, so we used the number of runs with 116 solution occurrences as a secondary measure of algorithm quality.

In Tables 1 and 2 we give results for CGP with fitness functions as in Eqs. 7 and 8 respectively, for different genotype sizes and mutation probabilities. The first value in each column represents the average value over all runs and the second value, in brackets, represents the number of obtained 116 nonlinearity solutions over all runs (higher is better for both values). The results for both fitness versions with GP for various tree depths are given in Table 3.

As it can be seen from the tables, CGP outperforms GA and GP quite easily. It should be noted that many additional GA and GP combinations were already experimented with in our previous research that are not shown here, which exhibit the same or worse performance than the configurations used in this work. Thus, we concentrate on the CGP efficiency which has not been previously investigated.

In addition, we can compare CGP variants with the weighted fitness and two-stage fitness. In Fig. 1 we plot the one-stage and two-stage fitness data shown in Tables 1 and 2 as a scatter graph showing the average fitness versus the genotype length for all mutation rates. We also show the number of nonlinearity 116 solutions found in both cases.

The results for the weighted fitness outperform two-stage fitness in many cases. This is a surprising result as a two-stage fitness is often used in CGP, ever since it was first described [7,14]. It implies that more work should be done

Table 1. Results for Eq. 7 and CGP.

Genotype/p_m	1	3	5	7	9	11	13
100	101.58 (0)	105.78 (1)	100.9 (0)	105.52 (0)	105.94 (2)	105.68 (1)	104.58 (0)
300	110.86 (2)	110.62 (14)	111.22 (13)	111.5 (16)	109.98 (12)	112.12 (16)	111.36 (10)
500	111.26 (11)	112.94 (20)	113.04 (24)	113.5 (24)	114.18 (25)	113.16 (21)	112.42 (20)
700	112.92 (15)	112.7 (23)	113.24 (26)	113.76(27)	113.98 (29)	113.54 (29)	113.16 (30)
900	110.72 (11)	114.38(31)	114.16 (28)	114.48 (31)	114.28 (30)	114.32 (31)	114.7 (34)
1 100	112.4 (10)	114.28 (29)	114.82 (35)	114.56 (33)	114.14 (27)	114.44 (34)	114.74 (36)
1 300	112.76 (12)	114.38 (30)	114.76 (35)	114.3 (30)	114.3 (32)	114.98 (37)	114.58 (34)
1 500	112.58 (12)	114.56 (34)	114.58 (33)	115.08 (40)	114.44 (35)	114.96 (37)	115.16 (39)
1 700	112.88 (15)	113.96 (27)	114.8 (35)	113.7 (29)	114.2 (32)	113.94 (29)	115.12 (40)
1 900	112.52 (12)	114.12 (31)	114.32 (33)	114.48 (31)	114.8 (36)	114.22 (29)	113.38 (25)

Table 2. Results for Eq. 8 and CGP.

Genotype/p_m	1	3	5	7	9	11	13
100	94.16 (0)	96.8 (2)	92.96 (0)	96.32 (0)	94 (1)	99.76 (0)	96.32 (0)
300	108.28(0)	108.00 (8)	107.6 (3)	109.68 (9)	102.56 (6)	104.72 (7)	107.36 (6)
500	106.64(1)	110.8 (7)	108.92 (7)	110.4 (6)	110.64 (13)	107.28 (9)	109.84 (9)
700	111.92 (5)	109.96 (11)	111.6 (15)	110.64 (15)	110.68 (9)	111.52 (14)	110.48 (7)
900	110.8 (5)	112.32 (13)	112.76 (20)	112.08 (17)	112.72 (17)	110.96 (15)	112.92 (16)
1 100	111.64 (8)	112.96 (17)	112.96 (19)	113.36 (17)	111.84 (11)	111.40 (13)	112.72 (12)
1 300	110.88 (2)	112.84 (19)	113.28 (17)	112.96 (20)	111.72 (12)	112.48 (13)	112.56 (12)
1 500	111.48 (2)	112.48 (9)	112.20 (13)	113.60 (20)	113.12 (19)	112.76 (14)	112.52 (11)
1 700	112.16 (8)	111.6 (15)	112.88 (15)	111.88 (17)	112.92 (16)	113.04 (20)	113.20 (17)
1 900	111.0 (5)	112.96 (15)	112.76 (17)	112.6 (14)	112.64 (23)	112.8 (15)	112.36 (10)

Table 3. Results for GP.

Tree depth	5	7	9	11	13
Eq. 7	112.13 (1)	112.2 (2)	111.36(0)	111.64 (0)	111.22 (0)
Eq. 8	112.13 (1)	112 (0)	111.76 (1)	111.72 (0)	111.58 (0)

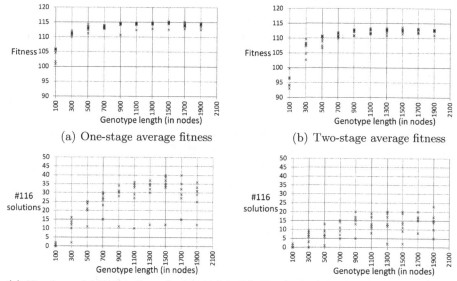

(a) One-stage average fitness

(b) Two-stage average fitness

(c) Number of 116 solutions found with one-stage average fitness

(d) Number of 116 solutions found with two-stage average fitness

Fig. 1. Comparative results for one-stage 1(a) and two-stage 1(b) fitness functions showing average fitness achieved and the number of 116 solutions found against genotype length for all mutation rates.

on a variety of problems to establish the relative merits of the two approaches. In addition in [12] it was suggested that optimal mutation rates should decrease as genotype length increases. However, the results here indicate that for the cryptographic problem studied this is not the case. Indeed fairly high mutation rates produced the best results. This is also surprising and merits further investigation.

When adding the resilience constraint to the fitness function, we observe that all Boolean functions within several generations obtain the resilience value of 0. This suggests that this condition is not hard enough objective to lead the search towards very high nonlinearity values in different parts of search space when compared with fitness functions without that objective.

When considering the average number of active nodes for CGP we give numbers for the best set of parameters and both fitness functions in Table 4. Notice that we selected the best algorithm on the basis of the number of achieved 116 nonlinearity values. In the case that two algorithms have the same number of 116 values, then we consider the average value as the second criteria.

We carried out longer runs of 10 million evaluations for the best combinations of CGP parameters considering the total number of obtained 116 nonlinearity values. We do not give similar comparison for GP since it is much slower and from that perspective is not competitive with CGP for such large number of evaluations. Figure 2 shows a boxplot comparison of best parameter

Table 4. Average number of active nodes.

Algorithm	CGP, Eq. 7	CGP, Eq. 8	CGP, Eq. 7, long run	CGP, Eq. 8, long run
Genotype, p_m	1 700, 13	1 900, 9	1 700, 13	1 900, 9
Value	84.24	76.62	81.76	86.33

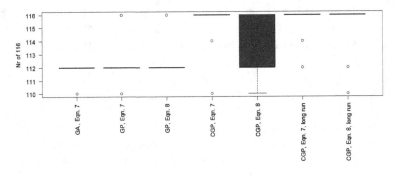

Fig. 2. Boxplot comparison of the most successful algorithms.

combinations for GA, GP and CGP with fitness functions Eqs. 7 and 8. Furthermore, we present best parameter combinations for CGP with 10 million evaluations. Note that the same parameter combinations for CGP are presented in Table 4.

5   Conclusion and Future Work

This paper describes an application of GA, GP and CGP in an evolution of cryptography relevant Boolean functions. The main contribution is the application of CGP, whose efficiency has not been previously investigated for this problem, and a comparison with two other methods. The results show that CGP is able to produce results that are clearly better than previous approaches, and is at the same time a valid choice for this kind of problem. Additionally, the described optimization problem may be considered a viable candidate as a benchmark problem for GP-related algorithms, both for its hardness as well as its real-world applicability.

Acknowledgments. This work was supported in part by the Technology Foundation STW (project 12624 - SIDES), The Netherlands Organization for Scientific Research NWO (project ProFIL 628.001.007) and the ICT COST action IC1204 TRUDEVICE.

References

1. Braeken, A.: Cryptographic Properties of Boolean Functions and S-Boxes. Ph.D. thesis, Katholieke Universiteit Leuven (2006)
2. Burnett, L.D.: Heuristic optimization of boolean functions and substitution boxes for cryptography. Ph.D. thesis, Queensland University of Technology (2005)

3. Cid, C., Kiyomoto, S., Kurihara, J.: The RAKAPOSHI stream cipher. In: Qing, S., Mitchell, C.J., Wang, G. (eds.) ICICS 2009. LNCS, vol. 5927, pp. 32–46. Springer, Heidelberg (2009)

4. Clark, J.A., Jacob, J.L., Stepney, S., Maitra, S., Millan, W.L.: Evolving boolean functions satisfying multiple criteria. In: Menezes, A., Sarkar, P. (eds.) INDOCRYPT 2002. LNCS, vol. 2551, pp. 246–259. Springer, Heidelberg (2002)

5. Crama, Y., Hammer, P.L.: Boolean Models and Methods in Mathematics, Computer Science, and Engineering, 1st edn. Cambridge University Press, New York (2010)

6. Goossens, K.: Automated creation and selection of cryptographic primitives. Master's thesis, Katholieke Universiteit Leuven (2005)

7. Kalganova, T., Miller, J.F.: Evolving more efficient digital circuits by allowing circuit layout evolution and multi-objective fitness. In: Proceedings NASA/DoD Workshop on Evolvable Hardware, pp. 54–63. IEEE Computer Society (1999)

8. Katz, J., Lindell, Y.: Introduction to Modern Cryptography. Chapman and Hall/CRC, Boca Raton (2008)

9. McLaughlin, J., Clark, J.A.: Evolving balanced boolean functions with optimal resistance to algebraic and fast algebraic attacks, maximal algebraic degree, and very high nonlinearity. Cryptology ePrint Archive, Report 2013/011 (2013). http://eprint.iacr.org/

10. Millan, W.L., Clark, A.J., Dawson, E.: Heuristic design of cryptographically strong balanced boolean functions. In: Nyberg, K. (ed.) EUROCRYPT 1998. LNCS, vol. 1403, pp. 489–499. Springer, Heidelberg (1998)

11. Millan, W., Fuller, J., Dawson, E.: New concepts in evolutionary search for boolean functions in cryptology. Computat. Intell. 20(3), 463–474 (2004)

12. Miller, J., Smith, S.: Redundancy and computational efficiency in cartesian genetic programming. IEEE Trans. Evol. Comput. 10(2), 167–174 (2006)

13. Miller, J.F.: An empirical study of the efficiency of learning boolean functions using a cartesian genetic programming approach. In: Banzhaf, W., Daida, J.M., Eiben, A.E., Garzon, M.H., Honavar, V., Jakiela, M.J., Smith, R.E. (eds.) GECCO, pp. 1135–1142. Morgan Kaufmann (1999)

14. Miller, J.F. (ed.): Cartesian Genetic Programming. Natural Computing Series. Springer, Heidelberg (2011)

15. Miller, J.F., Thomson, P.: Cartesian genetic programming. In: Poli, R., Banzhaf, W., Langdon, W.B., Miller, J., Nordin, P., Fogarty, T.C. (eds.) EuroGP 2000. LNCS, vol. 1802, pp. 121–132. Springer, Heidelberg (2000)

16. Miller, J.F., Harding, S.L.: Cartesian Genetic Programming. In: Proceedings of the 10th Annual Conference Companion on Genetic and Evolutionary Computation, GECCO 2008, pp. 2701–2726. ACM, New York (2008)

17. Nyberg, K.: Perfect nonlinear S-Boxes. In: Davies, D.W. (ed.) EUROCRYPT 1991. LNCS, vol. 547, pp. 378–386. Springer, Heidelberg (1991)

18. Picek, S., Batina, L., Jakobovic, D.: Evolving DPA-resistant boolean functions. In: Bartz-Beielstein, T., Branke, J., Filipič, B., Smith, J. (eds.) PPSN 2014. LNCS, vol. 8672, pp. 812–821. Springer, Heidelberg (2014)

19. Picek, S., Jakobovic, D., Golub, M.: Evolving cryptographically sound boolean functions. In: GECCO (Companion), pp. 191–192 (2013)

20. Picek, S., Marchiori, E., Batina, L., Jakobovic, D.: Combining evolutionary computation and algebraic constructions to find cryptography-relevant boolean functions. In: Bartz-Beielstein, T., Branke, J., Filipič, B., Smith, J. (eds.) PPSN 2014. LNCS, vol. 8672, pp. 822–831. Springer, Heidelberg (2014)

21. Prouff, E.: DPA Attacks and S-Boxes. In: Gilbert, H., Handschuh, H. (eds.) FSE 2005. LNCS, vol. 3557, pp. 424–441. Springer, Heidelberg (2005). http://www.iacr.org/cryptodb/archive/2005/FSE/3172/3172.pdf
22. Sarkar, P., Maitra, S.: Nonlinearity Bounds and Constructions of Resilient Boolean Functions. In: Bellare, M. (ed.) CRYPTO 2000. LNCS, vol. 1880, pp. 515–532. Springer, Heidelberg (2000)

Templar – A Framework
for Template-Method Hyper-Heuristics

Jerry Swan and Nathan Burles[✉]

Department of Computer Science,
University of York, York YO10 5GH, UK
jerry.swan@cs.york.ac.uk, nathan.burles@york.ac.uk

Abstract. In this work we introduce Templar, a software framework for customising algorithms via the generative technique of template-method hyper-heuristics. We first discuss the need for such an approach, presenting Quicksort as an example. We provide a functional definition of template-method hyper-heuristics, describe how this is implemented by Templar, and show how Templar may be invoked using simple client-code. Finally, we describe experiments using Templar to define a 'hyper-quicksort' with the aim of reducing power consumption— the results demonstrate that the generated algorithm has significantly improved performance on the test set.

Keywords: Genetic programming · Generative hyper-heuristics · Template method · Energy profiling · Reduced power consumption · Quicksort

1 Introduction

Despite two decades of research, scalability remains an issue for program synthesis via metaheuristics. Although there have been some recent promising results [20], for many problems generative approaches such as Genetic Programming (GP) [9] still work best at the scale of expressions. A greater degree of explicit structure can be imposed by Grammatical Evolution [19], although this approach is known to suffer from a lack of locality [18]. By contrast, human ingenuity has already provided a vast repertoire of specialized algorithms, usually with known asymptotic behaviour. Given these limitations of scale, how can we best use generative approaches to improve upon human-designed algorithms?

One motivating observation is that the performance of many well-known algorithms arises in practice from the inclusion of heuristically-informed decision points and/or *ad hoc* boundary conditions (e.g. the recursion depth at which Musser's widely-used sorting algorithm 'Introsort' switches from Quicksort to Heapsort [15]). The method of hyper-heuristics [4] ('heuristics to select or generate heuristics') performs metaheuristic search over function spaces in which the functions are themselves heuristics. This leads naturally to the idea of a generative hyper-heuristic approach to algorithm design, in which we search for superior versions of the heuristically-informed parts of an algorithm. As described in a recent paper on 'Template Method hyper-heuristics' [27], rather than hope to synthesise a complete algorithm from the 'bottom up', we can instead use the

© Springer International Publishing Switzerland 2015
P. Machado et al. (Eds.): EuroGP 2015, LNCS 9025, pp. 205–216, 2015.
DOI: 10.1007/978-3-319-16501-1_17

```
DoubleArray
qsort ( arr  :  DoubleArray ,  pivotFn  :  DoubleArray  →  Double )  {
  Double  pivot  =  pivotFn ( arr  );
  // ^^^ pivotFn can be varied generatively
  return  qsort ( arr . filter ( < pivot ), pivotFn )
      ++ arr . filter ( == pivot )
      ++ qsort ( arr . filter ( > pivot ), pivotFn );
}
```

Listing 1. Quicksort with variant pivot function

'Template Method' Design Pattern [6] and provide an algorithm skeleton (the template) that is parameterized by one or more 'variation points'. Each variation point is permitted to express a family of behaviours, whether constrained merely by the types of its function signature, or more strongly via 'design by contract'. By expressing an algorithmic framework in Template Method terms, we can then use generative techniques (in this case GP) to learn good implementations for the variant parts. By 'good', we mean 'biased towards the distribution to which the algorithm is exposed'. If our algorithms are metaheuristics, an important corollary is that they are not subject to the 'No Free Lunch' theorem [24], since the distribution over problem instances is biased away from uniform by the training set. This approach has been successfully demonstrated in the generation of more effective selection and mutation operators for Genetic Algorithms [25,26].

1.1 Quicksort - A Motivating Example

Although Quicksort [7] is of course a generic algorithm (i.e. it can be defined over any partially-ordered type), for simplicity of exposition we consider it to operate on arrays of floating point values, denoted by DoubleArray. Quicksort performance is well-known to be dependent on the choice of pivot, which we can therefore consider as a variation point for the algorithm. The pivot-function can be taken to have signature: pivotFn : DoubleArray → Double, i.e. it returns the choice of pivot value for a given input array. Listing 1 gives the pseudocode for a version of Quicksort that takes the pivot function as an additional argument. By specifying our algorithm in this fashion, we can generate a version that best meets some training criterion specified in client-code, such as robustness against pathological inputs (e.g. hardening against denial-of-service attacks), reduced power consumption, etc.

In the remainder of this article we describe TEMPLAR, a generic Java[TM] framework for Template Method hyper-heuristics, and show how it can be used for rapid prototyping. The outline of the article is as follows: in Sect. 2, we give a functional definition of Template Method hyper-heuristics and describe the corresponding TEMPLAR implementation. Starting from elementary examples, in Sect. 3 we show how TEMPLAR is configured and invoked from client-code. In Sect. 4, we detail an experiment with 'hyper-quicksort' to demonstrate the utility of this approach, giving conclusions in Sect. 5.

```
@FunctionalInterface
// ^ tells Java this can be treated as a lambda function
interface Fun1<Arg, Result> {
    Result apply(Arg arg);
}


@FunctionalInterface
interface Fun2<Arg1, Arg2, Result> {
    Result apply(Arg1 arg1, Arg2 arg2);
}

// We can now use functions as parameters
// and return values:
Fun1<A, C> compose(Fun1<A, B> f, Fun1<B, C> g) {
    return (A x) -> g.apply(f.apply(x));
}
```

Listing 2. Higher-order functions in Java

2   A Functional Framework for Template Method Hyper-Heuristics

For an algorithm with function signature $I \to O$, Template Method hyper-heuristics can be described as follows:

1. A list of *variation points* describing the parts of the algorithm to be automatically generated:

$$\text{VP} : (I_1 \to O_1) \times (I_2 \to O_2) \times \ldots \times (I_n \to O_n)$$

2. An *algorithm template* expressing the algorithm skeleton. The template produces a customized version of the algorithm from automatically-generated implementations of the variation points:

$$\text{Template} : \text{VP} \to (I \to O)$$

3. A *loss function* to evaluate the customized algorithm on supplied training and testing sets as a function of the difference between actual and expected outputs:

$$\text{LossFn} : (O \times O) \to V$$

4. An *algorithm factory* that searches the space of variation points to produce an optimized version of the algorithm:

$$\text{Factory} : \text{VP} \times \text{Template} \times \text{LossFn} \to (I \to O)$$

Despite the success of applications such as [5], the vast majority of hyper-heuristic publications concern selective hyper-heuristics—there are far fewer on generative approaches. Hoos [8] discusses automated algorithm improvement,

```
interface AlgTemplate<I, O> {
  public Fun1<I, O> makeAlg(ProgramList programs);
}

class AlgFactory<I, O> {
  AlgFactory(GPConfig[] variationPointConfigs,
    AlgTemplate<I, O> template) { ... }

  ProgramList run(FitnessCases<I, O> cases,
    LossFn<O> lossFn) { ... }
}
```

Listing 3. Core TEMPLAR classes

and the benefits this can bring—such as removing the menial work involved in manually experimenting with new algorithms and the improved performance of the resulting algorithms. Unfortunately generative hyper-heuristics are laborious to implement on a per-case basis, but also nontrivial to generalize. There are several reasons for the latter: firstly, generation of the variant programs is typically implemented via GP and is invoked repeatedly by the Factory in the process of the hyper-level search. Unfortunately, popular GP implementations such as ECJ [12] and PushGP [21] prefer to be the 'top' of the system (not least because of their 'configuration file' based approach) and hence attempting to use them for generative hyper-heuristics is not a simple matter. Secondly, the fitness of each variation point depends on the others, and a generic implementation of the complex 'wiring diagram' of dependencies is likely to be offputting to many experimenters. There has recently been some interesting related work using an algorithm configuration tool [10] to play an analogous role to the grammar in Grammatical Evolution [19]. This has been successfully used to automatically generate local search heuristics [13].

3   The TEMPLAR Framework

Functional programming makes intensive use of higher-order functions, i.e. functions which can accept (and significantly, return) other functions. Listing 2 shows how higher-order functions can be defined in Java. As of Java 8, equivalents of Fun1,Fun2 are supported natively as java. util .Function1,Function2 and Lambda functions can be expressed in the concise syntax we use in the program listings[1].

The core TEMPLAR classes are given in Listing 3. AlgFactory defines the process of hyper-heuristic search for program variants (supplied subclasses offer Iterative Improvement or Genetic Algorithms), whereas the end-user must subclass AlgTemplate in order to generate a new algorithm from a ProgramList containing a generated program for each of the variation points. At the top of Listing 4 is IdentityTemplate, the simplest possible subclass of AlgTemplate. It has

---

[1] docs.oracle.com/javase/tutorial/java/javaOO/lambdaexpressions.html.

```
class IdentityTemplate implements AlgTemplate<Double,
Double> {
  public Fun1<Double, Double> makeAlg(ProgramList progs) {
    // Wrap the VP in a function:
    return (Double arg) -> progs.get(0).execute(arg);
  }
}

class CompositionTemplate implements AlgTemplate<Int, String>
  {
  Fun1<Int, String> makeAlg(ProgramList progs) {
    Fun1<Int, Double> f = (Int arg) ->
      progs.get(0).execute(arg);
    Fun1<Double, String> g = (Double arg) ->
      progs.get(1).execute(arg);
    // this template composes the two variant programs:
    return compose(f, g);
  }
}
```

Listing 4. Simple AlgTemplate examples

no actual algorithm skeleton, i.e. it merely wraps the generated program of its (sole) variation point inside a function and returns it. This is therefore equivalent to a 'standard' (i.e. non-template) generative approach. Of slightly greater utility is CompositionTemplate, in which the algorithm skeleton is the composition of two generated variants f and g to give $f(g(x))$. As can be seen in Listings 4 and 5, using TEMPLAR requires that the end user do only the following:

1. Define an AlgTemplate subclass as described above.
2. Configure GP for each variation point (Listing 5).
3. Invoke TEMPLAR on user-supplied training and testing sets (Listing 5).

The GPConfig class contains all the information (function set, population size, mutation and crossover rates and operators, etc.) required to generate programs for each variation point. In Listing 5, the RationalFunctionConfiguration ($\{+, -, *, \%\}$) built-in to TEMPLAR is used. The lossFn parameter determines the fitness of algorithm variants as a function of the difference between actual and expected outputs—here root mean square error (RMS) is used.

In terms of computational expressiveness, TEMPLAR is equivalent to approaches such the Grammatical Evolution or the automated configuration tool *irace* mentioned above. Aside from the issue of lack of locality in such approaches [18], the main advantage we claim for TEMPLAR is that it is more programatically integrated: unlike grammar configuration files or parsed EBNF grammar strings, the entities manipulated by TEMPLAR are all first-class objects and the validity of the resulting program can be enforced to the extent that Java's type-system permits. This approach also brings some other benefits when

```
public static void main(String[] args) {
   // Configure GP for each variation point
   AlgTemplate<Double, Double> template =
      new IdentityTemplate();
   GPConfig[] vpConfigs = new GPConfig[] {
      new RationalFunctionConfig() };
   LossFn<Double> lossFn = new RMSLossFn<Double>();

   // Set-up training and testing sets:
   FitnessCases trainingSet = ...
   FitnessCases testSet = ...

   // Invoke TEMPLAR:
   ProgramList bestVPs = Templar.run(template, vpConfigs,
      trainingSet, testSet, lossFn);
   println("best VPs:" + bestVPs);
}
```

Listing 5. Configuring and running TEMPLAR

working with non-trivial datatypes. For example, the algorithm described above operates on values of type Double. For more sophisticated algorithms, it is desirable for the generated programs to operate directly on user-defined datatypes (e.g. BitString, Timetable, RoutePlan, AntTrail etc.). However, the manual creation of GP nodes for function sets on such custom representations is tedious. Following [11], a FunctionSetGenerator utility is provided by TEMPLAR, using reflection to automatically build a function set from the methods defined on *any* Java object.

4   Hyper-Quicksort

By following the steps described above, it is a simple matter to create hyper-heuristic versions of any algorithm. In this section, we describe how to create a 'hyper-quicksort'. Listing 6 gives the complete client code for this. Java is an unnecessarily verbose language and unlike languages such as C++ which can reduce syntactic clutter by using typedef to create a type alias, there is no explicit support in Java for this. Something similar can, however, be achieved by creating an appropriately named subclass, as has been done here with PivotFn.

It is well-known that Quicksort does not perform well on certain pathological distributions (e.g. nearly-sorted or reverse-sorted input) [14]. To demonstrate the effectiveness of our approach, we used as input a 'pipeorgan' distribution, in which the values in the input array increase monotonically until some randomly-specified index, then decrease monotonically. Quicksort is known to behave poorly against data drawn from this distribution, so the case study we present could be considered as a simple example of 'hardening' software against a denial-of-service attack. An example is given in Fig. 1.

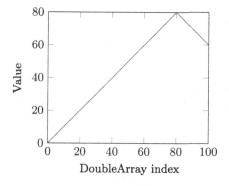

Fig. 1. Example of 'pipeorgan' training set for hyper-quicksort

4.1   Optimising for Energy Reduction

Measuring the power consumption of a computer can be performed using a number of methods, such as reading the data from an uninterruptible power supply, or the use of an electronic watt-meter. Recently, software-based alternatives have become available that use power models in conjunction with timings and system information such as CPU utilisation in order to provide reasonably accurate estimations [2]. These hardware and software tools can be very useful for informing users of their power consumption, or as a course-grained overview of energy used by an application. Unfortunately their precision and accuracy are too low to be suitable for use in automated software improvement—competing algorithms would need to be run an inordinate number of times to obtain a single measurement, essentially making the hyper-heuristic intractable. Although software-based tools can only provide estimates of the power consumed, the important factor is relative consistency between competing solutions. The use of a more accurate software measure is thus acceptable, such as the Wattch [3] and JALEN [16] tools. Wattch is a cycle-level simulator that has been used successfully with GP (e.g. [23]), however it requires a parameterised model of the processor and does not support Java. JALEN is a more recent alternative that targets Java, and can calculate an estimate of the power consumption by monitoring the execution time alongside system resources such as processor utilisation. Due to its simplicity and target language, in this work we have chosen to use JALEN to determine the fitness of competing algorithms.

4.2   Experimentation and Results

The experimental setup was as follows: the GP metaheuristic was configured with a population size of 100, 200 generations per run, an initial tree depth of 2 and a max tree depth of 4. These values were determined empirically as a reasonable trade-off between solution quality and execution time of the hyper-heuristic. All other GP parameters and operators were as the EpochX 1.4 defaults. An iterative-improvement hyper-heuristic was used on top of a GP metaheuristic, with a stopping condition of 100 iterations. The training set contained 70

```
// 1. Define an AlgTemplate subclass:

// 'typedef' for PivotFn to increase readability:
@FunctionalInterface
abstract class PivotFn
extends Fun2<DoubleArray, Int, Double> {
  Double apply(DoubleArray a, Int recursionDepth);
}

class QuicksortTemplate
implements AlgTemplate<DoubleArray, Int> {
  Fun1<DoubleArray, Int> makeAlg(ProgramList progs) -> {
    PivotFn pivotFn = (DoubleArray a, Int recursionDepth) -> {
      int progResult = programs.get(0).execute(a.size(),
          recursionDepth);
      int numSamples = Math.min(Math.abs(progResult),
          a.size());
      return median(randomSample(a, numSamples));
    };
    return (DoubleArray arg) -> Quicksort.sort(arg, pivotFn);
  }
}

Int quicksort(DoubleArray a, PivotFn pivotFn);
// ^ instrumented to return fitness (e.g. power consumed)

// 2. Configure GP to generate pivotFn VP:
List<Var> vars = {Var(''size''), Var(''recursionDepth'')};
List<Node> funcSet = {IfFn(), LessFn(), AddFn(), ...};
GPParams params = ... // crossover, selection, etc
GPConfig vpConfigs = {new GPConfig(funcSet, vars, params)};

// 3. Invoke TEMPLAR
AlgTemplate<Double, Double> template =
    new QuicksortTemplate();
FitnessCases trainingSet = ...
FitnessCases testSet = ...
Templar.trainAndTest(template, vpConfigs, trainingSet,
    testSet, new RMSLossFn<Double>());
```

Listing 6. Client-code for Hyper-quicksort

cases, where each case consisted of 100 arrays to be sorted, each with size 100. The function to be minimized was the energy used to perform the Quicksort, using a modified version of the JALEN tool[2].

---

[2] In order to provide the highest possible accuracy, JALEN was modified to run from within the Java Virtual Machine, rather than as an external Java Agent.

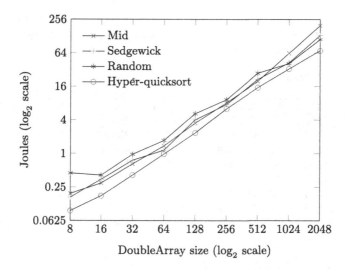

Fig. 2. Energy required to sort 1000 'pipeorgan' arrays using various pivot functions

As described above, the pivot function is the sole variation point, and the input variables for the generated programs are 'array size' and 'recursion depth'. The template for the evolved GP function is configured as follows: the Oracular pivot value of an array of values is its median. Here, we use GP to generate the function GP(arraySize, recursionDepth) → numSamplePoints and take as pivot the median of numSamplePoints randomly-chosen array elements. Although Quicksort is defined on anything with a partial order, note that this particular method only works when the array values are numeric. A number of different functions were generated by different runs of the GP, however commonly-observed amongst the fitter functions was recursionDepth. This suggests that as the recursion depth increases (and array size decreases), performance is improved by increasing the number of samples.

In Fig. 2, the performance is compared with three well-known pivot functions: 'Middle index', 'Random index', and 'Sedgewick' (the latter returning the median of the first, middle, and last elements). In this testing, JALEN was used to calculate the energy required to sort 1000 arrays of varying lengths using each of the pivot functions. This test was performed 100 times, and Fig. 2 presents the mean results. Hyper-quicksort can be seen to outperform all of the alternatives. In order to verify that the results are significant, we used the non-parametric Mann-Whitney U-test [1]. For each array length, $2^3 - 2^{11}$, we ran the U-test comparing the set of results obtained using hyper-quicksort to each of the alternative pivots in turn. As shown in Table 1, in all of the cases it was clear that the distributions of the results were significantly different, with $p < 0.05$.

Having demonstrated significance, it is important that the effect size is also measured [1]. For this we used Vargha and Delaney's non-parametric $\hat{A}_{12}$ statistic [22] for each of the sets of results above. This test returns the probability that using algorithm A provides higher values than using algorithm B. We wish

Table 1. Energy (J) required to sort 1000 'pipeorgan' arrays using various pivot functions, as well as the p-values (p) and effect size measures (e) comparing hyper-quicksort to the alternative pivot functions.

Array size	Middle index			Sedgewick			Hyper-quicksort (J)
	J	p	e	J	p	e	
8	0.191	7.46e-32	0.981	0.163	8.37e-32	0.980	0.094
16	0.296	1.20e-30	0.971	0.345	1.25e-31	0.979	0.173
32	0.651	8.13e-32	0.980	0.757	7.25e-32	0.981	0.410
64	1.366	4.80e-33	0.990	1.145	1.68e-30	0.970	0.976
128	3.505	4.80e-33	0.990	4.034	4.14e-33	0.991	2.341
256	8.175	4.14e-33	0.991	7.646	3.41e-32	0.983	6.387
512	19.777	4.33e-34	0.998	21.391	3.62e-34	0.999	15.268
1024	62.961	2.52e-34	1.000	42.508	6.44e-33	0.989	33.012
2048	198.438	2.52e-34	1.000	132.663	2.52e-34	1.000	70.234

Array size	Random index		
	J	p	e
8	0.446	8.37e-32	0.980
16	0.410	8.37e-32	0.980
32	0.967	4.80e-33	0.990
64	1.708	4.80e-33	0.990
128	5.221	2.52e-34	1.000
256	9.269	8.87e-34	0.996
512	27.685	2.52e-34	1.000
1024	41.245	3.61e-32	0.983
2048	111.894	3.47e-33	0.991

to minimize the energy used, and so in each case algorithm B is hyper-quicksort, and algorithm A the alternative—the results of the $\hat{A}_{12}$ statistic therefore give the probability that using the alternative pivot function will use more energy than hyper-quicksort. Vargha and Delaney suggest the following guidelines for interpreting the effect size: 0.5 indicates that there is no benefit to either algorithm; 0.56 indicates a small difference; 0.64 indicates a medium difference; and 0.71 indicates a large difference. As would be expected, there is some range in the effect sizes across the various array sizes and alternative pivot functions—from 0.970 to 1.000. The results are presented in full in Table 1, and show definitively that the Quicksort variant generated by TEMPLAR provides improved performance compared to common pivot functions, with respect to the energy required to sort families of data drawn from a pathological 'pipeorgan' distribution.

5   Conclusion and Future Work

We have introduced TEMPLAR, a framework that supports 'Template Method Hyper-heuristics' [27], by which any algorithm can be parameterized by a collection of variant subroutines to be generated automatically. By judicious choice of subroutines (and their associated function signatures), the framework can make effective use of Genetic Programming to tune the algorithm to some target distribution. Since the supporting algorithm skeleton can be arbitrarily complex, this allows greater scalability than can be achieved by the naïve application of Genetic Programming. In implementation terms, TEMPLAR makes creation of generative hyper-heuristics a more procedural matter of GP parameter tuning. We described how to create a 'Hyper-quicksort' and showed the effectiveness of the approach by optimizing for power consumption. Regarding future work, TEMPLAR currently uses EpochX [17] as the GP backend. It would also be desirable to support PushGP and ECJ as alternatives. This may be difficult because of their 'configuration file' based-approach, but if successful would allow direct comparison of their performance as a generative hyper-heuristic mechanism.

Although this approach has proven successful, the imprecision of the power measurement approach used in our experiments imposes an undesirable constraint on experimentation—algorithms must be run numerous times within a single measurement. This greatly increases the time required to run the hyper-heuristic. In future work it would be advantageous if a bytecode or opcode power model was developed for use with an execution trace. Such an implementation would allow highly precise modelling of power consumption, reducing the time required for optimizations and potentially improving results.

References

1. Arcuri, A., Briand, L.: A hitchhiker's guide to statistical tests for assessing randomized algorithms in software engineering. Softw. Test. Verif. Reliab. **24**(3), 219–250 (2014)
2. Bekaroo, G., Bokhoree, C., Pattinson, C.: Power measurement of computers: analysis of the effectiveness of the software based approach. Int. J. Emerg. Technol. Adv. Eng. **4**(5), 755–762 (2014)
3. Brooks, D., Tiwari, V., Martonosi, M.: Wattch: a framework for architectural-level power analysis and optimizations. In: 27th Annual International Symposium on Computer Architecture, ISCA 2000, pp. 83–94. ACM (2000)
4. Burke, E.K., Hyde, M.R., Kendall, G., Ochoa, G., Özcan, E., Woodward, J.R.: A classification of hyper-heuristic approaches. In: Gendreau, M., Potvin, J.Y. (eds.) Handbook of Metaheuristics. International Series in Operations Research and Management Science, vol. 146, pp. 449–468. Springer, New York (2010)
5. Burke, E.K., Hyde, M.R., Kendall, G., Woodward, J.R.: Automating the packing heuristic design process with genetic programming. Evol. Comput. **20**(1), 63–89 (2012)
6. Gamma, E., Helm, R., Johnson, R., Vlissides, J.: Design Patterns: Elements of Reusable Object-Oriented Software. Addison-Wesley, Boston (1995)
7. Hoare, C.A.R.: Algorithm 64: Quicksort. Commun. ACM **4**(7), 321 (1961)
8. Hoos, H.H.: Computer-aided design of high-performance algorithms. Technical report TR-2008-16, University of British Columbia, Department of Computer Science (2008)
9. Koza, J.R.: Genetic Programming: On the Programming of Computers by Means of Natural Selection. MIT Press, Cambridge (1992)

10. López-Ibáñez, M., Dubois-Lacoste, J., Stützle, T., Birattari, M.: The irace package, iterated racing for automatic algorithm configuration. Technical report TR/IRIDIA/2011-004, IRIDIA, Université Libre de Bruxelles, Belgium (2011)
11. Lucas, S.: Exploiting reflection in object oriented genetic programming. In: Keijzer, M., O'Reilly, U.-M., Lucas, S., Costa, E., Soule, T. (eds.) EuroGP 2004. LNCS, vol. 3003, pp. 369–378. Springer, Heidelberg (2004)
12. Luke, S.: The ECJ Owner's Manual, zeroth edn., online version 0.2 edn., October 2010. http://www.cs.gmu.edu/~eclab/projects/ecj
13. Mascia, F., López-Ibáñez, M., Dubois-Lacoste, J., Stützle, T.: Grammar-based generation of stochastic local search heuristics through automatic algorithm configuration tools. Comput. Oper. Res. **51**, 190–199 (2014)
14. McIlroy, M.D.: A killer adversary for quicksort. Softw. Pract. Exp. **29**(4), 341–344 (1999)
15. Musser, D.R.: Introspective sorting and selection algorithms. Softw. Pract. Exp. **27**(8), 983–993 (1997)
16. Noureddine, A., Bourdon, A., Rouvoy, R., Seinturier, L.: Runtime monitoring of software energy hotspots. In: 27th IEEE/ACM International Conference on Automated Software Engineering 2012, pp. 160–169. IEEE (2012)
17. Otero, F., Castle, T., Johnson, C.: EpochX: genetic programming in Java with statistics and event monitoring. In: GECCO 2012, pp. 93–100. ACM, New York (2012)
18. Rothlauf, F., Oetzel, M.: On the locality of grammatical evolution. In: Collet, P., Tomassini, M., Ebner, M., Gustafson, S., Ekárt, A. (eds.) EuroGP 2006. LNCS, vol. 3905, pp. 320–330. Springer, Heidelberg (2006)
19. Ryan, C., Collins, J.J., O'Neill, M.: Grammatical evolution: evolving programs for an arbitrary language. In: Banzhaf, W., Poli, R., Schoenauer, M., Fogarty, T.C. (eds.) EuroGP 1998. LNCS, vol. 1391, pp. 83–96. Springer, Heidelberg (1998)
20. Spector, L., Harrington, K., Helmuth, T.: Tag-based modularity in tree-based genetic programming. In: GECCO 2012, pp. 815–822. ACM, New York (2012)
21. Spector, L., Klein, J., Keijzer, M.: The Push3 execution stack and the evolution of control. In: GECCO 2005, pp. 1689–1696. ACM, New York (2005)
22. Vargha, A., Delaney, H.D.: A critique and improvement of the CL common language effect size statistics of McGraw and Wong. J. Educ. Behav. Stat. **25**(2), 101–132 (2000)
23. White, D.R.: Genetic programming for low-resource systems. Ph.D. thesis, University of York (2009)
24. Woodward, J.R., Swan, J.: Why classifying search algorithms is essential. In: International Conference on Progress in Informatics and Computing (2010)
25. Woodward, J.R., Swan, J.: Automatically designing selection heuristics. In: GECCO 2011, pp. 583–590. ACM, New York (2011)
26. Woodward, J.R., Swan, J.: The automatic generation of mutation operators for genetic algorithms. In: GECCO 2012, pp. 67–74. ACM, New York (2012)
27. Woodward, J.R., Swan, J.: Template method hyper-heuristics. In: GECCO 2014, pp. 1437–1438. ACM, New York (2014)

Circuit Approximation
Using Single- and Multi-objective Cartesian GP

Zdenek Vasicek and Lukas Sekanina[✉]

Faculty of Information Technology, IT4Innovations Centre of Excellence,
Brno University of Technology, Božetěchova 2, 612 66 Brno, Czech Republic
{vasicek,sekanina}@fit.vutbr.cz

Abstract. In this paper, the approximate circuit design problem is formulated as a multi-objective optimization problem in which the circuit error and power consumption are conflicting design objectives. We compare multi-objective and single-objective Cartesian genetic programming in the task of parallel adder and multiplier approximation. It is analyzed how the setting of the methods, formulating the problem as multi-objective or single-objective, and constraining the execution time can influence the quality of results. One of the conclusions is that the multi-objective approach is useful if the number of allowed evaluations is low. When more time is available, the single-objective approach becomes more efficient.

Keywords: Genetic programming · Cartesian genetic programming · Evolutionary design · Approximate computing · Approximate circuits · Multi-objective approach

1 Introduction

Approximate computing is a promising approach for the design of energy efficient computer-based systems (see detailed motivation and survey in [1,5]). It exploits the fact that many applications are error resilient which means that their users are willing to accept less than perfect solutions, simply because the inaccuracies in the output are not recognizable, or they are well justified under some circumstances. Multimedia applications, search, classification, prediction and recognition tasks are typical domains for approximate computing. Approximations can be introduced at the circuit, component, architecture, software, operating system or system's level. In some cases, the degree of approximation can be adapted during system's deployment. Because of the nature of evolutionary algorithms which evolve target systems by introducing small changes into existing structures, it seems that evolutionary computing could be an efficient method to approximate (i.e. purposely modify) circuit designs [10,11].

From the designer's perspective, a reasonable trade-off is sought between the accuracy and power consumption. (Alternatively, the accuracy can be traded for the speed of operation in some applications.) The approximate circuit design

© Springer International Publishing Switzerland 2015
P. Machado et al. (Eds.): EuroGP 2015, LNCS 9025, pp. 217–229, 2015.
DOI: 10.1007/978-3-319-16501-1_18

problem can be formulated as a multi-objective design and optimization problem in which the accuracy and power consumption are conflicting design objectives. A good approximate circuit design tool should provide a set of solutions which exhibit various trade-offs among key circuit parameters, in particular, the accuracy and power consumption. These solutions should, in an idealized case, perfectly match the so-called Pareto optimal front [2]. Current approximation tools (such as [9,12,13]) typically solve this problem by multiple executions of approximation engines in order to obtain a set of various solutions. With respect to given constraints and specification, the designer finally selects one of the compromises to be implemented on a chip.

The approximate circuit design is a computationally demanding process which involves generating and comparing many circuit designs. In order to justify this computation time, the resulting circuit should really represent a good compromise between the target objectives. The maximum number of circuits that is allowed to be generated and evaluated thus becomes the main constraint for approximation engines.

The goal of this paper is to compare multi-objective and single-objective versions of Cartesian genetic programming (CGP) [8] in the task of combinational circuit approximation. The reasons for using an advanced evolutionary approach (contrasted to a greedy search used in the state of the art tools [9]) are that the population-based approach suits well in finding multiple solutions and its niche-preservation methods can be exploited to discover diverse solutions [2].

The methodology presented in this paper uses the following principles: (1) the single- and multi-objective search methods are compared under various constraints on the execution time because the design time is one of the crucial factors determining applicability of a design method; (2) the key circuit parameters (area and delay) are estimated during the optimization process while the resulting approximate circuits are implemented using a standard design flow and compared with their accurate counterparts. It has to be noted that performing a fair comparison of various approximation algorithms is not trivial in practice because different teams have the access to different test circuits (some of them are proprietary) and fabrication technologies (correct power estimation depends on a particular fabrication process).

Section 2.1 surveys relevant methods developed to approximate circuit designs. Section 2.2 is devoted to the principles of multi-objective optimization. The proposed single-objective and multi-objective approximate circuit design methods are introduced in Sect. 3. Experimental results are presented in Sect. 4. Conclusions are given in Sect. 5.

2   Related Work

2.1   Approximate Computing

In approximate computing systems, the accuracy (or quality) of the output is traded for improvements in power consumption or performance. This is possible because many applications are intrinsically error resilient and users are willing in many cases to accept less than perfect performance or quality. Approximations are currently applied at all system's levels [5]. We will solely focus on approximate circuits in this paper.

Initial approaches to the functional approximation have been based on a manual identification of subcircuits that should be approximated, for example, in adders and multipliers [7]. However, the manual approach is not efficient and scalable. Later, several systematic automated methodologies have been proposed [9,10,12,13].

For example, ABACUS creates an abstract synthesis tree from the input behavioral description and then applies various operators to it using an iterative stochastic greedy approach [9]. Candidate designs are evaluated in terms of accuracy, power consumption and area using a single objective optimization algorithm. The objectives are combined together in a weight function. The Pareto front is obtained from multiple runs of the search algorithm (only about 50 candidate circuits are generated in each run [9]).

The aforementioned methods try to approximate the Pareto optimal front by either combining more design objectives in a single objective search (ABACUS) or executing the approximation algorithm with one fixed criterion (e.g. the error is constant) and optimizing for another one (minimizing power consumption). However, in many cases, the resulting solutions do not cover the whole Pareto front and the design alternatives are centered around a few dominant designs. These methods employ the standard design flow to construct and evaluate every candidate circuit, which is very time consuming. On the other hand, the circuit parameters obtained are very close to real ones.

Systematic methods based on the evolutionary design paradigm consider the approximate circuit design problem as a search problem. It was exploited in [10,11] that power consumption is often highly correlated with occupied resources and the evolutionary design is capable of constructing partly working solutions even if sufficient resources (required for finding a fully functional solution) are not available. The user then obtains, in multiple runs of CGP, a set of approximate combinational circuits, each of which typically exhibits different trade-off between the accuracy and the number of gates. Delay was not addressed in papers [10,11].

2.2 Multi-objective Optimization

In general, the multi-objective optimization problem can be defined in the following form:

$$
\begin{aligned}
\text{optimize: } & f_m(\boldsymbol{x}), & m &= 1, 2, ..., M \\
\text{subject to: } & g_j(\boldsymbol{x}) \geq 0 & j &= 1, 2, ..., J \\
& h_k(\boldsymbol{x}) = 0 & k &= 1, 2, ..., K
\end{aligned}
\tag{1}
$$

where $\boldsymbol{x} = (x_1, x_2, \ldots, x_n)$ is a vector representing the solution consisting of n decision variables. The objective functions are denoted $f_1, \ldots, f_M$. Some of these functions have to be minimized, others have to be maximized. Functions $g_j(\boldsymbol{x})$

and $h_k(\boldsymbol{x})$ define the optimization constrains and thus determine the space of feasible solutions.

In order to compare two solutions, Pareto-dominance relations are employed [2]: Solution $\boldsymbol{x}^{(1)}$ *dominates* another solution $\boldsymbol{x}^{(2)}$ if the following conditions are satisfied: (i.) The solution $\boldsymbol{x}^{(1)}$ is no worse than $\boldsymbol{x}^{(2)}$ in all objectives. (ii.) The solution $\boldsymbol{x}^{(1)}$ is strictly better than $\boldsymbol{x}^{(2)}$ in at least one objective.

The result of the multi-objective optimization is no longer a single solution, but a set of solutions. In a set of solutions P, a non-dominated subset of solutions P' contains those solutions that are not dominated by any member of P. The non-dominated subset of all possible solutions is called Pareto-optimal set (front). The members of this subset are optimal solutions to the multi-objective optimization problem. The ultimate goal of any multi-objective optimization algorithm is to find all solutions which belong to the Pareto-optimal front. In practice, the goal is to find a set of solutions as close as possible and as diverse as possible with respect to the Pareto-optimal front.

A straight forward approach to the multi-objective optimization is converting the multi-objective problem to a single objective one using a weight function $\sum w_i f_i$, where w_i is the weight of the i-th objective. Because a single run of the optimizer which uses the sum yields only one solution, multiple runs are needed for obtaining various trade-offs. The proper setting of weights w_i is not an easy task and is usually based on the user intuition. Another limitation of the weight function lies in the fact that certain Pareto-optimal solutions are not reachable in the case of nonconvex objective space [2]. Since it is difficult to detect whether the resulting objective space is nonconvex, the weight function has to be applied with caution.

In order to precisely approximate the whole Pareto-optimal front and obtain various diverse non-dominate solutions in a single run of an optimizer, truly multi-objective evolutionary algorithms have been introduced, for example, non-dominated sorting genetic algorithm (NSGA-II). Contrasted to the single-objective optimization algorithms, they internally sort individuals according to the dominance relation, build archives of non-dominating solutions, and ensure population diversity to avoid converging to a single solution. In the context of evolutionary design of (exact) circuits, multi-objective CGP has been applied in [6,8].

3   The Proposed Search Methods

The proposed approach is based on Cartesian genetic programming [8] and its multi-objective extension utilizing the NSGA-II [3].

3.1   Circuit Representation

A candidate circuit is modeled by means of a directed acyclic graph whose nodes (gates) are organized in n_c columns and n_r rows. The circuit has n_i primary inputs and n_o primary outputs. Each node input can be connected either to the output of a node placed in previous l columns or to one of the primary circuit inputs, where l is one of CGP parameters.

A candidate solution consisting of two-input nodes is represented in the chromosome by $n_r \cdot n_c$ triplets (x_1, x_2, ψ) determining for each processing node its function ψ ($\psi \in \Gamma$), and addresses of nodes x_1 and x_2 which its inputs are connected to. The last part of the chromosome contains n_o integers specifying either the nodes where the primary outputs are connected to or logic constants ('0' and '1') which can directly be connected to the primary output. While the chromosome size is constant for a given product $n_r \cdot n_c$, the phenotype size is variable and measured as the number of used nodes (gates).

3.2 Single-Objective Search

The initial population of CGP is created either randomly or seeded by available circuits. Candidate circuits are evaluated using the fitness function. If a multi-objective optimization is conducted, there are several fitness functions formulated, each of them reflecting to what extent a given circuit parameter (accuracy, area, delay etc.) satisfies the specification.

When multiple-objectives are aggregated to a single fitness value (e.g. using the weight function), we speak about a single-objective optimization. Each member of the population then receives one fitness value and the highest-scored idividual becomes a new parent of the next population.

The offspring circuits are created from the parent using mutation, which is the only operator used in CGP. The mutation modifies h randomly selected genes (integers) of the parent circuit. CGP usually employs a $1 + \lambda$ search strategy. The evolution is terminated when a predefined number of generations is exhausted or a suitable solution is discovered.

3.3 Multi-objective Search

In the multi-objective algorithm, the $1 + \lambda$ search strategy is replaced by procedures of NSGA-II which implement non-dominated sorting of the population (non-dominated solutions are emphasized) and diversity preservation mechanisms (less crowded points of the search space are promoted) – details can be found in [3]. Here, the population consists of λ_{MO} individuals. The non-dominated sorting algorithm of NSGA-II was modified in such a way that when all components of the fitness score of a parent and its offspring remain unchanged, the offspring is classed as dominating the parent, and is therefore ranked higher than the parent. Moreover, the maximum allowed error E_{max} (which the designer is going to observe and accept in the resulting Pareto fronts) is defined as a constraint in our algorithm. In order to optimize the error (inaccuracy), area and delay, three fitness functions (all to be minimized) will be constructed. If fitness $f_{error} > E_{max}$, the solution is considered as unacceptable.

3.4   Methodology

Because many candidate circuits will be generated and evaluated in the course of evolution, it is intractable to precisely calculate power consumption and other circuit parameters for each of them. Hence we will only calculate the error and estimate the area and delay. As power consumption is highly correlated with the area (which can be assumed for certain technology nodes), it is particularly important to find good compromises between the area and error. At the end of the evolutionary optimization, selected approximate circuits will be implemented using a standard design flow and compared with their fully functional versions.

In order to estimate parameters of a given circuit, the area and delay are calculated using the parameters defined in the liberty timing file available for a given semiconductor technology. This file gives the area, timing and power-relevant parameters of each cell (gate).

Delay t_d of a cell c_i is modeled as a function of its input transition time t_s and capacitive load c_l on the output of the cell, i.e. $t_d(c_i) = f(t_s^{c_i}, c_l^{c_i})$. Delay of circuit C is determined as delay of the longest path:

$$Delay(C) = \max_{\forall p \in \text{path}} \sum_{c_i \in p} t_d(c_i).$$

The capacitive load on the circuit outputs is chosen to be equal to the input capacitance of an inverter cell. The transition time on circuit inputs corresponds to the transition time on the output of an inverter cell.

The area of circuit C is calculated as the sum of areas of all cells c_i involved in the circuit:

$$Area(C) = \sum_{c_i \in C} area(c_i).$$

Various error criteria can be utilized to evaluate the quality of an approximate arithmetic circuit. The average error magnitude $E_{avg}(C)$ is employed in our case. This metric is defined as the sum of absolute differences in magnitude between the original and approximate circuits averaged over all inputs:

$$E_{avg}(C) = \frac{\sum_{\forall i} |Y(C_{orig}, i) - Y(C, i)|}{2^{2w}} \tag{2}$$

where $Y(C_{orig}, i)$ denotes the output value of the fully functional circuit for the input vector i, $Y(C, i)$ denotes the output value of approximate circuit C and w specifies the bit-width.

Multi- as well as single-objective algorithm is seeded with a fully functional version of an arithmetic circuit. In both cases the user is supposed to define E_{max}. However, the interpretation of E_{max} is different. In the case of multi-objective optimization, solutions with the error greater than E_{max} are unacceptable; the remaining solutions are considered during the Pareto front construction. In the case of single-objective optimization, the evolutionary algorithm is used to find a solution showing the error as close as possible to E_i. In order to construct Pareto front, the single-objective algorithm has to be executed multiple times with E_i increasing from a small error to E_{max} in several steps.

When a solution showing the error E_i has to be discovered, the single-objective algorithm works in two stages. The aim of the first stage is to produce a circuit with an error as close as possible to the required level E_i regardless the other optimization criteria. To achieve this objective, the fitness value $fitness_{L1}$ is calculated as the relative absolute difference from the required error level and the goal is to minimize this difference, i.e.

$$fitness_{L1}(C) = \frac{|E_{avg}(C) - E_i|}{E_i}$$

If the required error is obtained ($fitness_{L1} < 0.01$), the algorithm continues by the second stage. In this stage, additional optimization objectives are taken into account and the error E_i serves as a constraint which guarantees that the average error value is kept as close as possible to the target error level. Each objective is normalized to the interval $< 0, 1 >$ and weighted with a weight w_e, w_a or w_d ($w_e + w_a + w_d = 1$). Then,

$$fitness_{L2}(C) = \begin{cases} w_e E'_{avg}(C)+ \\ w_a Area'(C)+ \\ w_d Delay'(C), & \text{if } fitness_{L1} < 0.01 \\ \infty, & \text{otherwise,} \end{cases}$$

where the apostrophe denotes a normalized value with respect to the original, fully functional circuit.

4 Results

4.1 Experimental Setup

The single-objective (SO) and multi-objective (MO) algorithms based on CGP are evaluated in the task of 4-bit and 8-bit adder and multiplier approximation. E_{max} is chosen to be 2.5 % of the maximum average error, where the maximum average error is $(2^w - 1)^2$ for the multiplier and $2(2^w - 1)$ for the adder. While the multi-objective algorithm is executed with E_{max}, the single-objective algorithm is executed SO_{run} times; one run for one error level from 0 % to 2.5 %. In both cases, CGP was initialized by fully functional circuits. We compared three sets of weights for the single-optimization algorithm $w_e/w_a/w_d = \{(0.8, 0.12, 0.08), (0.5, 0.38, 0.12), (0.12, 0.5, 0.38)\}$ (inspired in [9]).

In both approaches we used the following CGP parameters: $h = 5$, $l = n_c = N_g$, $n_r = 1$, $n_i = 2w$, $n_o = 2w$ for w-bit multiplier and $n_o = w + 1$ for w-bit adder, where N_g is the number of gates of the original fully functional circuit. In both cases, $5 \cdot 10^3$ evaluations (fitness calls) were allowed, corresponding to 100 generations of the multi-objective algorithm ($\lambda_{MO} = 50$). In the case of the single-objective algorithm, 50 generations are produced for each of 20 error levels ($\lambda_{SO} = 5$, $SO_{run} = 20$). This number of evaluations is very low from the perspective of evolutionary circuit design; however, this number is still much higher than in conventional methods [9].

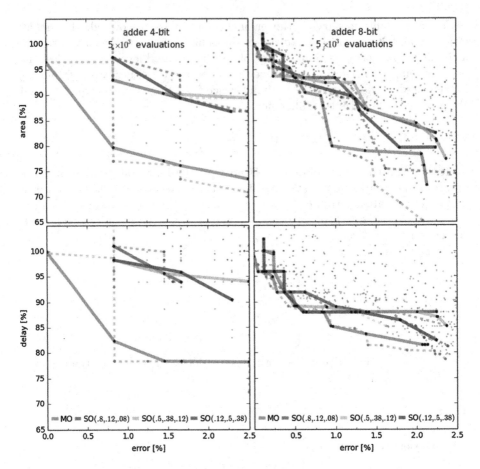

Fig. 1. Adders: Performance of single- and multi-objective methods with respect to original fully functional circuits (0 % error).

The experiments were conducted for I3T25 technology (0.35 μm digital process). The following cells (and thus functions in Γ) are considered: *and, or, xor, nand, nor, xnor, buf, inv*, with corresponding relative areas 1.333, 1.333, 2, 1, 1, 2, 1.333, and 0.667.

4.2   Comparison of Single- and Multi-objective Search

Figures 1 and 2 show the resulting parameters of all circuits as dots in two 2D plots (area vs. error and delay vs. error). These figures contain results from 25 independent runs of the algorithms for each scenario. The 3D Pareto front (projected to two 2D graphs) is interpolated using solid lines for each investigated scenario. Other Pareto fronts (dashed lines) are constructed in such a way that one objective (either area or delay) is ignored. These (dashed) Pareto fronts represent better compromises because the problem is simplified. One can observe that MO performs better than SO and the role of weight setting is less important.

In another experiment, we investigated whether increasing the number evaluations to $500 \cdot 10^3$ can improve the quality of results. Figure 3 shows that with decreasing the acceptable error, the results produced by SO are better than those from MO. The weight of the area in the fitness function (SO) becomes more important and its unsuitable setting can influence the result by 20 %. The SO approach exploits the fact that the error is fixed and the overall effort can be put into minimizing the area and delay. On the other hand, MO has to cover the whole Pareto front and the available time seems to be insufficient to compete with SO.

In order to further investigate the computation requirements, we analyzed the quality of resulting solutions with respect to the number of allowed evaluations (generations) in Fig. 4. In the case of MO, the progress of evolution is negligible after $50 \cdot 10^3$ evaluations. SO is capable of improving the quality until $250 \cdot 10^3$ evaluations are spent on average.

The computational requirements of the multi-objective algorithm are slightly higher than for the single-objective method (457 vs. 491 evaluations per second for the 8-bit adder and 240 vs. 284 evaluations per second for the 8-bit multiplier).

Table 1. Parameters of the approximate 8-bit adders and multipliers. The Impr. columns give the improvement w.r.t. the original circuits.

	Error	estimated				professional tool					
		Delay	Impr.	Rel.Area	Impr.	Delay	Impr.	Area	Impr.	Power	Impr.
	[%]	[ns]	[%]	[–]	[%]	[ns]	[%]	[μm^2]	[%]	[μW]	[%]
adder	0.0	2.8	–	99	–	2.7	–	4759	–	208	–
	0.6	2.5	12	51	49	2.2	19	2460	49	104	50
	1.2	2.1	27	32	68	1.9	29	1622	66	71	66
	1.9	1.7	38	28	72	1.6	40	1374	72	55	74
	2.5	1.8	36	20	80	1.6	39	978	80	42	80
multiplier	0.0	13.1	–	495	–	12.1	–	24245	–	1367	–
	0.6	8.7	34	175	65	8.0	35	8480	66	409	71
	1.3	6.3	52	118	77	5.6	54	5424	78	233	83
	1.9	5.4	59	92	82	5.1	58	4513	82	164	88
	2.5	4.5	66	64	87	4.3	65	3118	88	106	93

4.3 Results of Synthesis

In order to validate the presented results, we implemented selected circuits using a standard design flow. The original circuits and selected circuits obtained by CGP were converted into a netlist, and after synthesis, placement and routing (Cadence Encounter RTL Compiler), we compared parameters of resulting circuits with the estimated values used during the evolution. Table 1 shows that

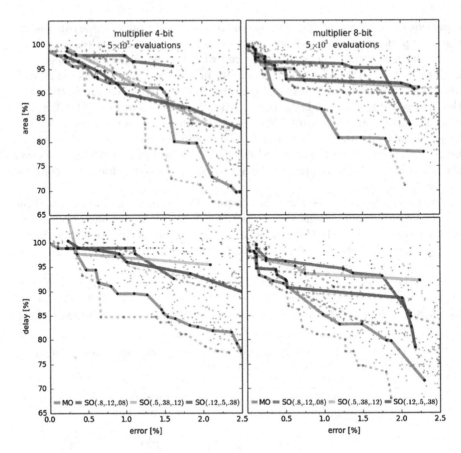

Fig. 2. Multipliers: Performance of single- and multi-objective methods with respect to original fully functional circuits (0 % error).

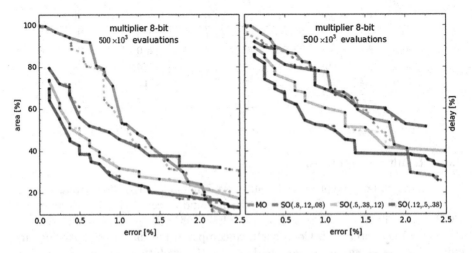

Fig. 3. Resulting Pareto fronts when $500 \cdot 10^3$ evaluations are allowed

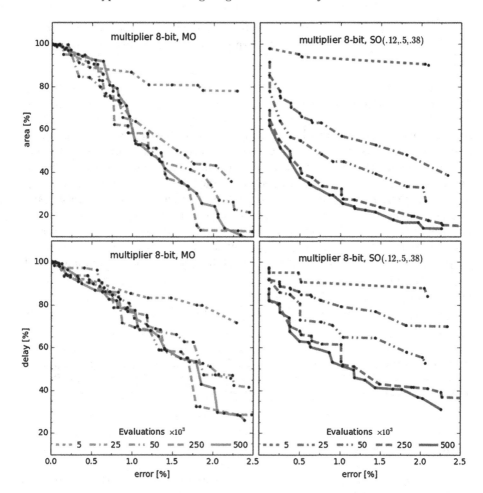

Fig. 4. Resulting Pareto fronts with respect to the number of evaluations

our estimated values are almost perfectly matched with the results of the professional tool (see Impr. columns). The area is correlated with power consumption under the investigated scenario.

A direct and fair comparison with some results from the literature is difficult for several reasons: implementations of methods such as SASIMI and SALSA are not available; only some parameters of benchmark circuits reported in the literature are known (i.e. their implementations are not available); and results are given for different fabrication technology. However, one can observe that the proposed method led to 71 % power reduction with 0.6 % average error, which seems to be a good result for 8-bit multiplier in comparison with SASIMI [12]

(45 % power reduction with 0.5 % average error) and Gupta et al. [4] (35 % power reduction with 2.5 % average error), despite the fact that different technology was used.

5   Conclusions

In this paper, we proposed and compared two evolutionary approximation circuit design methods based on single- and multi-objective CGP. Contrasted to current approaches, in which every candidate circuit is implemented and evaluated by means of a professional design tool, candidate circuits' parameters are only quickly estimated in the optimization process. It allowed us to generate many more candidate designs than state-of-the art methods. It was shown that the multi-objective method is useful if the number of allowed evaluations is low. On the other hand, when more time is available, the single-objective method outperforms the multi-objective one. We validated key circuit parameters of selected approximate circuits by means of a professional design tool. By employing the advanced optimization algorithms and allowing more computation time, we obtained very good approximations of the Pareto optimal fronts for adders and multipliers.

Acknowledgments. This work was supported by the Czech science foundation project Advanced Methods for Evolutionary Design of Complex Digital Circuits 14-04197S. The authors would like to thank Jiri Petrlik for useful discussions on multi-objective evolutionary optimization.

References

1. Chakradhar, S.T., Raghunathan, A.: Best-effort computing: Re-thinking parallel software and hardware. In: Proceedings of the 47th Design Automation Conference - DAC, pp. 865–870. ACM (2010)
2. Deb, K.: Multi-Objective Optimization using Evolutionary Algorithms. Wiley, Chichester (2001)
3. Deb, K., Pratap, A., Agarwal, S., Meyarivan, T.: A fast and elitist multiobjective genetic algorithm: Nsga-ii. IEEE Trans. Evol. Comput. **6**(2), 182–197 (2002)
4. Gupta, V., Mohapatra, D., Raghunathan, A., Roy, K.: Low-power digital signal processing using approximate adders. IEEE Trans. CAD Integr. Circ. Syst. **32**(1), 124–137 (2013)
5. Han, J., Orshansky, M.: Approximate computing: An emerging paradigm for energy-efficient design. In: Proceedings of the 18th IEEE European Test Symposium, pp. 1–6. IEEE (2013)
6. Hilder, J., Walker, J., Tyrrell, A.: Use of a multi-objective fitness function to improve Cartesian genetic programming circuits. In: NASA/ESA Conference on Adaptive Hardware and Systems, pp. 179–185. IEEE (2010)
7. Kulkarni, P., Gupta, P., Ercegovac, M.D.: Trading accuracy for power in a multiplier architecture. J. Low Power Electron. **7**(4), 490–501 (2011)
8. Miller, J.F.: Cartesian Genetic Programming. Springer, Berlin (2011)

9. Nepal, K., Li, Y., Bahar, R.I., Reda, S.: Abacus: A technique for automated behavioral synthesis of approximate computing circuits. In: Proceedings of the Design, Automation and Test in Europe, pp. 1–6. DATE 2014, EDA Consortium (2014)
10. Sekanina, L., Vasicek, Z.: Approximate circuits by means of evolvable hardware. In: IEEE Int. Conf. on Evolvable Systems, SSCI-ICES. pp. 21–28. IEEE CIS (2013)
11. Vasicek, Z., Sekanina, L.: Evolutionary approach to approximate digital circuits design. IEEE Tran. on Evolutionary Computation, pp. 1–13 (2015 to appear)
12. Venkataramani, S., Roy, K., Raghunathan, A.: Substitute-and-simplify: a unified design paradigm for approximate and quality configurable circuits. Design. Automation and Test in Europe, DATE 2013, pp. 1367–1372. EDA Consortium San Jose, CA, USA (2013)
13. Venkataramani, S., Sabne, A., Kozhikkottu, V.J., Roy, K., Raghunathan, A.: Salsa: systematic logic synthesis of approximate circuits. In: The 49th Annual Design Automation Conference 2012, DAC 2012, pp. 796–801. ACM (2012)

Author Index

Printed in the United States
By Bookmasters